GREEN CHEMISTRY AND GREEN ENGINEERING

Processing, Technologies, Properties, and Applications

GREEN CHEMISTRY AND GREEN ENGINEERING

Processing, Technologies, Properties, and Applications

Edited by

Shrikaant Kulkarni, PhD
Neha Kanwar Rawat, PhD
A. K. Haghi, PhD

First edition published 2021

Apple Academic Press Inc.
1265 Goldenrod Circle, NE,
Palm Bay, FL 32905 USA

4164 Lakeshore Road, Burlington,
ON, L7L 1A4 Canada

CRC Press
6000 Broken Sound Parkway NW,
Suite 300, Boca Raton, FL 33487-2742 USA

2 Park Square, Milton Park,
Abingdon, Oxon, OX14 4RN UK

© 2021 Apple Academic Press, Inc.

Apple Academic Press exclusively co-publishes with CRC Press, an imprint of Taylor & Francis Group, LLC

Reasonable efforts have been made to publish reliable data and information, but the authors, editors, and publisher cannot assume responsibility for the validity of all materials or the consequences of their use. The authors, editors, and publishers have attempted to trace the copyright holders of all material reproduced in this publication and apologize to copyright holders if permission to publish in this form has not been obtained. If any copyright material has not been acknowledged, please write and let us know so we may rectify in any future reprint.

Except as permitted under U.S. Copyright Law, no part of this book may be reprinted, reproduced, transmitted, or utilized in any form by any electronic, mechanical, or other means, now known or hereafter invented, including photocopying, microfilming, and recording, or in any information storage or retrieval system, without written permission from the publishers.

For permission to photocopy or use material electronically from this work, access www.copyright.com or contact the Copyright Clearance Center, Inc. (CCC), 222 Rosewood Drive, Danvers, MA 01923, 978-750-8400. For works that are not available on CCC please contact mpkbookspermissions@tandf.co.uk

Trademark notice: Product or corporate names may be trademarks or registered trademarks and are used only for identification and explanation without intent to infringe.

Library and Archives Canada Cataloguing in Publication

Title: Green chemistry and green engineering : processing, technologies, properties, and applications / edited by Shrikaant Kulkarni, PhD, Neha Kanwar Rawat, PhD, A.K. Haghi, PhD.

Names: Kulkarni, Shrikaant, editor. | Rawat, Neha Kanwar, editor. | Haghi, A. K., editor.

Description: Includes bibliographical references and index.

Identifiers: Canadiana (print) 20200308769 | Canadiana (ebook) 20200308807 | ISBN 9781771889001 (hardcover) | ISBN 9781003057895 (ebook)

Subjects: LCSH: Green chemistry. | LCSH: Sustainable engineering.

Classification: LCC TP155.2.E58 G74 2021 | DDC 660.028/6—dc23

Library of Congress Cataloging-in-Publication Data

Names: Kulkarni, Shrikaant, editor. | Rawat, Neha Kanwar, editor. | Haghi, A. K., editor.

Title: Green chemistry and green engineering : processing, technologies, properties, and applications / edited by Shrikaant Kulkarni, PHD, Neha Kanwar Rawat, PHD, A.K. Haghi, PHD.

Description: 1st edition. | Palm Bay : Apple Academic Press, 2021. | Includes bibliographical references and index. | Summary: "This interdisciplinary and accessible new volume presents a broad range of application-based green chemistry and engineering research. The goal of this research-oriented book is to familiarize readers with the integration of tools and spell out the approaches for green engineering of new processes as well as improving the environmental risks of existing processes. The expert authors discuss the myriad opportunities and the challenges facing green chemistry today in both its theoretical and practical implementation. The book expands upon green chemistry concepts with the latest research and new and innovative applications, providing both the breadth and depth researchers need. Topics include solar energy, electrospinning of bio-based polymeric nanofibers, biotransformation, engineered nanomaterials in environmental protection, and much more"-- Provided by publisher.

Identifiers: LCCN 2020036174 (print) | LCCN 2020036175 (ebook) | ISBN 9781771889001 (hardcover) | ISBN 9781003057895 (ebook)

Subjects: LCSH: Green chemistry. | Sustainable engineering.

Classification: LCC TP155.2.E58 G7274 2021 (print) | LCC TP155.2.E58 (ebook) | DDC 660.028/6--dc23

LC record available at https://lccn.loc.gov/2020036174

LC ebook record available at https://lccn.loc.gov/2020036175

ISBN: 978-1-77188-900-1 (hbk)
ISBN: 978-1-00305-789-5 (ebk)

About the Editors

Shrikaant Kulkarni, PhD
Assistant Professor, Vishwakarma Institute of Technology
Department of Chemical Engineering, India
E-mail: shrikaant.kulkarni@vit.edu

Shrikaant Kulkarni, PhD, has 37 years of teaching and research experience at both undergraduate and postgraduate levels. He has been teaching subjects such as engineering chemistry, green chemistry, analytical chemistry, catalysis, chemical engineering materials, and industrial organization and management. He has published over 100 research papers in national and international journals and conferences. He has authored 18 book chapters and has edited two books in green engineering and renewable materials. Another three books—on carbon nanotubes for green environment and carbon-based nanomaterials for energy storage and artificial intelligence for chemical sciences—are in the process of development. He has also co-authored four textbooks on chemistry. He is a reviewer and editorial board member of many international journals in the fields of green chemistry and analytical chemistry. He was invited by UNESCO to give a talk on "Green Chemistry Education for Sustainable Development" at the IUPAC international conference on green chemistry held in Bangkok, Thailand, in 2018. He is an esteemed member as a co-author on a team with sustainable development goals laid down by the United Nations. In addition, he was appointed as an esteemed jury member for the evaluation of business plans submitted by teams of students in the 13–18 age group from across the world for the Conrad challenge competition sponsored by NASA. He has been instrumental in formulating and coordinating RIO & COP programs dedicated to sustainable development at his institute. He is a resource person in materials science, analytical, polymer, and green and sustainable chemistry.

Neha Kanwar Rawat, PhD
Researcher in the Materials Science Division
CSIR-National Aerospace Laboratories, Bangalore, India
E-mail: neharawatjmi@gmail.com

Neha Kanwar Rawat, PhD, is a recipient of the prestigious DST Young Scientist Postdoctoral Fellowship and is presently a researcher in the Materials Science Division of CSIR-National Aerospace Laboratories, Bangalore,

India. She received her PhD in chemistry from Jamia Millia Islamia (A Central University), India. Her main interests include nanotechnology-nanostructured materials synthesis and characterization. The main focus comprises green chemistry, novel sustainable chemical processing of nano-conducting polymers/nanocomposites (NCs), conducting films, ceramics, silicones, matrices: epoxies, alkyds, polyurethanes, etc. She also pursues interests infusing new technology in areas include electrochemistry, organic-inorganic hybrid nanocomposites, and protective surface coatings for corrosion inhibition and MW shielding materials. She has published numerous peer-reviewed research articles in journals of high repute. Her contributions have led to many chapters in international books. She has worked on many prestigious research and academic fellowships in her career. She is a member of many groups, including the Royal Society of Chemistry and the American Chemical Society (USA), and is a life member of the Asian Polymer Association.

A. K. Haghi, PhD

Professor Emeritus of Engineering Sciences, Former Editor-in-Chief,
International Journal of Chemoinformatics and Chemical Engineering,
and Polymers Research Journal; Member,
Canadian Research and Development Center of Sciences and Culture,
Canada
E-mail: AKHaghi@gmail.com

A. K. Haghi, PhD, is the author and editor of over 200 books, as well as more than 1000 published papers in various journals and conference proceedings. Dr. Haghi has received several grants, consulted for a number of major corporations, and is a frequent speaker to national and international audiences. Since 1983, he has served as a professor at several universities. He is former Editor-in-Chief of the *International Journal of Chemoinformatics and Chemical Engineering* and *Polymers Research Journal* and is on the editorial boards of many international journals. He is also a member of the Canadian Research and Development Center of Sciences and Cultures (CRDCSC), Montreal, Quebec, Canada.

Contents

Contributors .... *ix*

Abbreviations .... *xi*

Preface .... *xvii*

1. **Electrospinning of Bio-Based Polymeric Nanofibers for Biomedical and Healthcare Applications** 1

    Rinky Ghosh, Veereshgouda Shekharagouda Naragund, and Neha Kanwar Rawat

2. **Solar Energy for Green Engineering Using a Multicomponent Absorption Power Cycle** 23

    Satchidanand R. Satpute, Nilesh A. Mali, and Sunil S. Bhagwat

3. **Characterization Techniques for Bio-Nanocomposites** 49

    Shrikaant Kulkarni

4. **Chemistry Matrices of Biotransformation** 83

    Shrikaant Kulkarni

5. **Sustainable Chemistry and Pharmacy** 109

    Hitesh V. Shahare and Shweta S. Gedam

6. **Magnetism, Polyoxometalates, Layered Materials, and Graphene** 123

    Francisco Torrens and Gloria Castellano

7. **Plasma, Photo/Radiochemical Reactions, and Relativity Theories** 135

    Francisco Torrens and Gloria Castellano

8. **Mesoporous, Graphene Composite, Li-Battery, Topology, and Periodicity** 149

    Francisco Torrens and Gloria Castellano

9. **Boom for Circular Economy and Creativity: Chemistry to Improve Life** 163

    Francisco Torrens and Gloria Castellano

10. **Application of Engineered Nanomaterials in Environmental Protection and the Visionary Future** 189

    Sukanchan Palit

viii

11. Chemistry and Sustainable Development...**211**

Francisco Torrens and Gloria Castellano

12. Manganese Porphyrins as Pro-Oxidants in High-Molar-Mass Hyaluronan Oxidative Degradation...**223**

Katarína Valachová, Peter Rapta, Ines Batinic-Haberle, and Ladislav Šoltés

13. Modified Magnetic Metal-Carbon Mesoscopic Composites with Bioactive Substances...**243**

V. I. Kodolov, I. N. Shabanova, V. V. Kodolova-Chukhonzeva, N. S. Terebova, Yu. V. Pershin, R. V. Mustakimov, and N. M. Pogudina

14. Pre-Treatment of Rusted Steel Surfaces with Phosphoric Acid Solutions...**255**

Victor M. M. Lobo and Artur J. M. Valente

15. Transport Phenomena of Electrolytes in Aqueous Solutions: Concepts, Approaches, and Techniques.................................**269**

Ana C. F. Ribeiro, Eduarda F. G. Azevedo, Ana Paula Couceiro Figueira, and Victor M. M. Lobo

Index.. *285*

Contributors

Eduarda F. G. Azevedo
Centro de Química, Department of Chemistry, University of Coimbra, 3004-535 Coimbra, Portugal

Ines Batinic-Haberle
Ines Batinic-Haberle, Department of Radiation Oncology-Cancer Biology,
Duke University Medical Center, Durham, USA

Sunil S. Bhagwat
Department of Chemical Engineering, Institute of Chemical Technology, Matunga, Mumbai, India

Gloria Castellano
Departamento de Ciencias Experimentales y Matemáticas, Facultad de Veterinaria y Ciencias
Experimentales, Universidad Católica de Valencia San Vicente Mártir, Guillem de Castro-94,
E-46001 València, Spain

Ana Paula Couceiro Figueira
Faculty of Psychology and Educational Sciences, University of Coimbra, 3004-535 Coimbra, Portugal

Shweta S. Gedam
Sandip Institute of Pharmaceutical Sciences, Mahiravani, Nasik, Maharashtra, India

Rinky Ghosh
Department of Materials Science and Engineering, Indian Institute of Technology Kanpur,
Kanpur – 208016, India

V. I. Kodolov
Scientific Head of BRHE Center of Chemical Physics and Mesoscopics, Professor,
M.T. Kalashnikov Izhevsk State Technical University, University in Izhevsk, Russia

V. V. Kodolova-Chukhonzeva
Scientific Head of BRHE Center of Chemical Physics and Mesoscopics, Professor,
M.T. Kalashnikov Izhevsk State Technical University, University in Izhevsk, Russia

Shrikaant Kulkarni
Assistant Professor, Vishwakarma Institute of Technology, Department of Chemical Engineering,
666, Upper Indira Nagar, Bibwewadi, Pune – 411037, India, Mobile: +91-9970663353,
E-mail: shrikaant.kulkarni@vit.edu

Victor M. M. Lobo
Centro de Química, Department of Chemistry, University of Coimbra, 3004-535 Coimbra, Portugal,
E-mail: vlobo@ci.uc.pt

Nilesh A. Mali
Chemical Engineering Division, National Chemical Laboratory, Pashan, Pune, India

R. V. Mustakimov
Postgraduate, BRHE Center of Chemical Physics and Mesoscopics, Izhevsk, Russia

Veereshgouda Shekharagouda Naragund
Materials Science Division, CSIR-National Aerospace Laboratories, Bangalore, India

Sukanchan Palit
Assistant Professor (Selection Grade), Department of Chemical Engineering,
University of Petroleum and Energy Studies, Energy Acres, Post-Office-Bidholi via Premnagar,
Dehradun – 248007, Uttarakhand, India; 43, Judges Bagan, Haridevpur (PO), Kolkata – 700082, India,
E-mails: sukanchan68@gmail.com, sukanchan92@gmail.com, sukanchanp@rediffmail.com

Yu. V. Pershin
Postgraduate, BRHE Center of Chemical Physics and Mesoscopics, Izhevsk, Russia

N. M. Pogudina
Researcher, BRHE Center of Chemical Physics and Mesoscopics, Izhevsk, Russia

Peter Rapta
Institute of Physical Chemistry and Chemical Physics, Slovak University of Technology in Bratislava,
Bratislava, Slovakia

Neha Kanwar Rawat
Senior Lecturer, Chemistry, Haryana Education Services, Haryana, India;
Materials Research Laboratory, Department of Chemistry, Jamia Millia Islamia, New Delhi, India,
E-mail: neharawatjmi@gmail.com

Ana C. F. Ribeiro
Centro de Química, Department of Chemistry, University of Coimbra, 3004-535 Coimbra, Portugal,
E-mail: anacfrib@ci.uc.pt

Satchidanand R. Satpute
Department of Chemical Engineering, Vishwakarma Institute of Technology, Bibwewadi, Pune, India,
E-mail: satchidanand.satpute@vit.edu

I. N. Shabanova
Professor, Udmurt Federal Research Center, Izhevsk, Russia

Hitesh V. Shahare
SNJB's Shriman Sureshdada Jain College of Pharmacy, Chandwad, Maharashtra, India,
E-mail: hiteshshahare1@rediffmail.com

Ladislav Šoltés
Center of Experimental Medicine, Institute of Experimental Pharmacology and Toxicology,
Slovak Academy of Sciences, Bratislava, Slovakia

N. S. Terebova
Doctor, Udmurt Federal Research Center, Izhevsk, Russia

Francisco Torrens
Institut Universitari de Ciència Molecular, Universitat de València, Edifici d'Instituts de Paterna,
P. O. Box – 22085, E-46071 València, Spain, E-mail: torrens@uv.es

Katarína Valachová
Center of Experimental Medicine, Institute of Experimental Pharmacology and Toxicology,
Slovak Academy of Sciences, Bratislava, Slovakia, E-mail: katarina.valachova@savba.sk

Artur J. M. Valente
CQC, Department of Chemistry, University of Coimbra, 3004-535 Coimbra, Portugal

Abbreviations

2D	two-dimensional
AA	ascorbic acid
AE	atom economy
AES	auger electron spectroscopy
AFM	atomic force microscopy
AMS	atomic mass spectrometry
APP	ammonium polyphosphate
AS	active site
ATP	adenosine tri phosphorus
ATR	attenuated total reflectance
BH	black hole
BWR	boiling water reactor
BZ	Brillouin zone
CA	chronoamperometric
CCC	countercurrent chromatography
CD	circular dichroism
CE	carbon efficiency
CH_2O	carbohydrate
CO_2	carbon dioxide
CPE	constant phase element
CRs	chain reactions
Cs	conclusions
CT	critical temperature
Cu-C MC	copper carbon mesoscopic composite
CV	cyclic voltamperometre
DLS	dynamic light scattering
DM	diamagnetism
DMA	dynamic mechanical analysis
EDS	energy dispersive x-ray spectroscopy
EELS	electron energy loss spectroscopy
EF	electric field
EIS	electrochemical impedance spectroscopy
EM	electromagnetic
EMY	effective mass yield

EPA	Environmental Protection Agency
EPR	electron paramagnetic resonance
ESI	electrospray ionization
ETH	Zürich Eidgenössische Technische Hochschule Zürich
FCT	Fundação para a Ciência e a Tecnologia
FLU	fluorescence
FM	ferromagnetic
FPF	fine particle fraction
FRs	free radicals
FTIR	Fourier transform infrared spectrometry
G	gaseous
GDH	glucose dehydrogenase
GPS	global positioning system
GR	graphene
GSK	Glaxo Smith Kline
GTR	general theory of relativity
GW	gravitational-wave
H	hypotheses
H_3PO_4	phosphoric acid
HA	hyaluronan
HA	hydroxyapatite
HCs	hydrocarbons
HF	Hartree-Fock
HF	high-frequency
HHDH	halohydrin dehalogenase
HILIC	hydrophilic interaction liquid chromatography
HN	hydroxynitrile product
HTs	high temperatures
ICP	inductively coupled plasma
ICP-AES	inductively coupled plasma-atomic emission spectroscopy
ICP-MS	inductively coupled plasma-mass spectroscopy
IG	ideal gas
IR	infrared
IS	interstellar space
ITC	isothermal titration calorimetry
JWST	James Webb space telescope
KQs	key questions
KRED	ketoreductase
L and RCPL	left and right-handed circularly polarized light

Abbreviations

L	liquid
LCA	life cycle analysis
LDHs	layered double hydroxides
LP	low pressure
LTs	low temperatures
MA	magnetic anisotropy
MAE	magnetic anisotropy effect
MCM	Mobil composition material
MDGC	multidimensional gas chromatography
MF	magnetic field
MFLs	MF force lines
MHD	magnetohydrodynamic
MI	mass intensity
MnPs	manganese porphyrins
MnSOD	manganese superoxide dismutase
MOFs	metal-organic frameworks
MP	mass productivity
MP	medium pressure
MS	mass spectrometry
MS-FTIR	mass spectrometry-Fourier-transform infrared spectroscopy
MSs	magnetic susceptibilities
MT	MF trap
MWs	microwaves
NAC	N-acetylcysteine
NAP	near ambient pressure
NCs	nanocomposites
NG	natural gas
NMR	nuclear magnetic resonance
NMs	nanomaterials
NS	nanoscale
ORC	organic Rankine cycles
OSs	oxidation states
P	plasma
PCS	photon correlation spectroscopy
PEO	poly(ethylene oxide)
PGA	polyglycolide
PLA	poly-lactic acid
PLGA	poly lactic-co-glycolic acid
PM	paramagnetism

PMs	porous materials
PMT	photomultiplier tube
POM	polyoxometalate
PPO	poly(propylene oxide)
Pt	platine
PTE	periodic table of the elements
PTs	phase transitions
PV	photovoltaic
QM	quantum mechanics
RCL	rust conversion layers
RE	renewable energy
Red-Ox	reduction-oxidation
RME	reaction mass efficiency
ROS/RNS	reactive oxygen and nitrogen species
RPES	x-ray photoelectron spectroscopy
RT	room temperature
S	solid
SBA	*Santa Barbara Amorphous*
SCF	self-consistent field
SCs	superconductors
SEM	scanning electron microscopy
SEs	side effects
SF	silk fibroin
SMs	standard models
SPM	scanning probe microscopy
SPR	surface plasmon resonance
SPRs	structure-property relationships
SQUID	superconducting quantum interference device
SSA	specific surface area
STEM	scanning transmission electron microscopy
STM	scanning tunneling microscopy
SVR	surface to volume ratio
SWNHs	single-wall nature of nanohorns
SWNTs	single-wall carbon nanotubes
TBG	twisted bilayer GR
TCD	tip-to-collector
TEM HP	transition electron microscopy with high permission
TEM	transmission electron microscopy
TEOS	tetraethylorthosilicate

TFs	thin films
TIP	thermodynamics of irreversible processes
TLC	thin-layer chromatography
TNT	trinitrotoluene
TP	transport properties
TQC	topological quantum chemistry
TTM	tris(triazolyl)methane
UHCs	unsaturated hydrocarbons
UV	ultraviolet
VIS	visible
VSM	vibrating sample magnetometer
WDXRF	wavelength dispersive X-ray fluorescence spectrometry
WWI	World War I
XPS	x-ray photoelectron spectroscopy
XRF	x-ray fluorescence spectrometry
ZnO	zinc oxide

Preface

As you are aware, we now live in an environmentally conscious era, in which sustainable materials and sustainable processing sequences are key considerations with regard to producing any product. This research-oriented book presents a complete picture of the current state-of-the-art in green chemistry and green engineering used in terms of the latest innovations, original methods, and safe products. Our main goal is to provide the necessary theoretical background and details on techniques, mechanisms, industrial applications, safety precautions, and environmental impacts. This research-oriented book is aimed at professionals from industry, academicians engaged in chemical engineering or natural product chemistry research, and graduate-level students.

Green chemistry refers to the study of the general methodology for the synthesis of chemicals in a benign and environmentally safe manner. Similarly, green engineering refers to the application of green chemistry on an industrial scale with the goal of designing processes, which minimizes waste and pollution. The practice of green engineering requires the integration of green chemistry concepts and systematic use of pollution prevention heuristics in design and operation together with risk assessment tools and life cycle analysis (LCA) tools.

Meanwhile, the goal of this book is to familiarize readers with the integration of these tools and present the approaches one needs to use for green engineering of new processes as well as improving the environmental risks of existing processes.

In this book, the authors and editors consider green chemistry and engineering principles, pollution prevention heuristics to be used in the design, environmental performance assessment, and life cycle assessment of processes, and prediction of the environmental fate of chemicals released into the environment.

The interest in the electrospinning process for the medicinal chemistry and nanotechnology field has escalated recently. This is because of simplicity in its design, cost-effectiveness, and interconnected porosity, better encapsulation of therapeutic agents, thermal consistency, higher loading capacity, and controlled release of drugs/gene carriers. Electrospun scaffolds recently have high behest for successful fabrications of scaffolds due to its significant endocytosis interactions with the cells and suitable degradation profile, tremendously exposed in the implants surgery and tissue regenerations.

Scaffolds of biocompatible nanofibers offer an efficient sustainable route for easy production of and facilitate the tailoring of surface properties by grafting and functionalization with bioactive agents. The biological studies have shown high potential in applications of biocompatible and biodegradable nanofiber scaffolds for wound dressing patches and drug delivery. Chapter 1 briefly discusses the electrospinning fundamental approaches to produced nanofibers scaffolds, biocompatible properties, drug-releasing rate kinetics, and its various applications in the interdisciplinary field of medicinal science.

Energy is a bottleneck for the development of modern civilization. Green energy becomes a way towards the sustainable future of human society. The increasing demand for energy and scarcity of the energy resources is directing us towards the innovative ways for the generation of power. Solar energy is abundantly available; hence, harnessing solar energy efficiently would be the most preferable option for power generation. This thought has given rise to new methodologies to study the thermodynamic systems. Established power generation uses a steam power plant where coal is used primarily as heat source works in the range of temperature 540–46°C and pressure of 150 bar–0.1 bar. Using solar easily steam can be produced up to 150°C. This low-temperature energy available couldn't be effectively used using traditional steam power plants. Chapter 2 deals with different efforts starting with traditional ways of producing power. The efficiencies pros and cons of various methods for power generation will be compared based on 1st law, 2nd law efficiencies, exergy efficiency, etc. The cumulative analysis of power generation relating to the temperature, pressure of the available heat sources will be carried out. The most efficient way to harness low-temperature energy using a multicomponent mixture will be presented with elaborate analysis with a unique platform of exergy relating to all different available methods and sources for power generation.

The characterization at different scales or thorough investigations of nano-composites, consisting of one or more inorganic particles and biomolecules, demands a host of techniques. It involves the exploration of bio-nanocomposite materials at the atomic level, elemental composition, extent, and the nature of the chemical bonds present. Moreover, it refers to the determination of size and structure, other than the identification of interactions between the inorganic and biological components of the system. Further, specific properties of the materials are examined not only because

Preface

xix

a material exhibits a specific function but also it gives the insight to probe the inorganic-biological interface. Conventional techniques too drawn from chemistry, material science, and biology have been employed studying these complex bio-nanocomposite materials. Thus an overview of key analytical methods is presented in Chapter 3, ranging from physical principles to practical aspects is taken which throws light upon various aspects of bio-nanocomposite materials.

Green and sustainable chemical manufacturing can be done using suitable catalysts as well as by employing meaningful green matrices for ascertaining the greenness of the processes and quantitation of reaction efficiency. Green chemistry matrices will also help in monitoring the kinetics of the reactions, e.g., atom efficiency and E factors are very widely recognized matrices across the board. If the green chemistry matrices are used in conjunction with life cycle assessment (LCA), it will provide a sound and concrete parameters for assessment of greenness and sustainability of spectrum of chemical processes. Catalysis like homogeneous, heterogeneous, organo-catalysis, and bio-catalysis will play an important role in order to evolve at sustainable technologies. Bio-catalysis, namely, has a comparative advantage in this regard in terms of mild reaction conditions with a catalyst which is biocompatible, biodegradable, and renewable with step economic, and highly selective processes reflecting upon better product quality, economical, and reduced waste generation. Moreover, the other benefits accrued include process intensification by virtue of the integration of bio-catalytic steps with cascading effect in processes. Bio-catalysis, will further the transition from a global economy relying upon non-renewable fossil feedstocks to a sustainable one which is renewable resources based driven by the greening of the synthetic routes or pathways.

Therefore, Chapter 4 highlights a discussion of a host of green chemistry matrices that are put into use for ensuring the greenness and sustainability in the manufacturing processes.

Green chemistry is today's need to minimize the environmental pollution by manmade materials and the processes, which are used to produce them. By application of these green principles involved in green chemistry, provides an opportunity for sustainable future and world development. The challenge in this is to face and maintain the green chemistry practice with their rules in the process. The use of these principles is now days extended from academic laboratories to industries. Green chemists used the green technique and its process design for the generation of biodegradable or recyclable products, to create sustainable raw materials, which help to prevent the waste. The growth of green chemistry will give positive benefits in various areas of business profit, culture, and protection of our earth, which is highlighted in Chapter 5.

It was pointed out the relation between the measurement that the two main molar magnetic susceptibilities (MSs) of graphite (-5×10^{-6} in the plane, -275×10^{-6} normal to the plane) and the theory of metallic conduction, which is shown by the graphite. Magnetic anisotropy (MA) and average magnetic susceptibility of graphite are reported to decay with the decaying size of the crystal grain. Graphite is included with metals because it shows metallic conductivity and the magnetic properties are, to a remarkable degree, similar to those found in some metals. The atomic magnetic susceptibility of graphite at room temperature (RT) is $\chi_A = 42\times10^{-6}$. Attention was drawn to the fact that graphite presents a large diamagnetic anisotropy. Transition metal dichalcogenides MX2 (M = Ti, V, Zr, Nb, Ta; oxidation states (OSs) 5, 4, X = S, Se, Te) are studied for superconductors (SCs). The periodic table of the elements (PTE) of MX2 for superconductors is presented in Chapter 6. The nanoscale (NS) is only compatible with the external layer(s) of the bulk material. The history of electronics begins with Ge, then passes to Si and, finally, to C (with oxidation states 4, –4). The periodic table of the elements of the history of electronics is presented in Chapter 6.

Plasma, photo/radiochemical reaction, and relativity theories discussed in Chapter 7. The quantum yield is the main characteristic on classifying photochemical reactions because its value varies within broad limits. A feature of photochemical reactions is that their rate does not practically depend on the temperature. Photochemical and radiochemical reactions are responsible for processes going on with a rise in Gibbs function of the system, and decay in the entropy of the universe, i.e., non-spontaneous processes, which aspect points to the alluring prospect of using the reactions for a number of syntheses. The vacuum is not the nothingness; it is a substance $\phi \neq 0$. If the vacuum is a substance, one can shake it; its vibrations are elementary particles: the Higgs. The standard models (SMs) of micro and macrocosmos burst. Einstein invented the general theory of relativity (GTR) and quantum mechanics (QM) wave-particle photon, but nobody knows how to unify them. In Einstein's blackboard, there is a lot to fill.

In Chapter 8, the importance of mesoporous, graphene composites, Li-battery, topology, and periodicity have been taken into consideration. The history of electronics began with germanium, passed to silicon, and finally, to carbon. The periodic table of the elements of the history of electronics was reported. Perhaps mesoporous silica-based materials will be the future

Preface

of electronics. Topological quantum chemistry (TQC) is a predictive theory of topological hands. Taking the ideas to their logical conclusion, a paradigm appears which applies not only to topological insulators but also to semimetals and band theory. The synthesis of symmetry and topology of localized orbitals and Bloch wave functions must enable a full understanding of noninteracting solids. The emphasis on the symmetry of localized orbitals opens up incorporating magnetic groups or interactions in the theory of topological materials. Predicted *d/f*-electron compounds must be analyzed by more accurate dynamical *mean-field theory* codes, which properly take into account interactions and can potentially give many new interacting materials with strong topology. Mimics of two-dimensional materials are possible in practice with the products outperforming the bulk materials, while control over stability can be governed by immobilization on solid surfaces. It was investigated how best to integrate experiments, computation, and theory. Science is seen from outside as experimental but another theoretical science exists, and classification is a part of theoretical science.

In Chapter 9, chemistry to improve people's life is discussed in detail. Creativity in science is the ability to get oneself into a fine mess. Chemophobia false myth: All that *smells* chemistry is bad. Positive meaning: *Between two persons, chemistry exists.* Solubility old rule: *Like dissolves like.* Research is to raise oneself questions and try to answer them. Technicians are the *why* but farmers are the *how*. New ethos: Spain–US Agreement. Better things for better living *via* chemistry. Not all fascist science is bad but excellent science requires social freedom. Practically, no historical rhetoric exists and, in Spanish-Civil-War Generation, totals rupture. Ingenuity and particular nature of chemistry for armor-plating itself. Tacit knowledge: intangible. Operation Paperclip: many German scientists recruited in post-Nazi Germany and taken to the US for government employment. Historical processes exist that require a long time. If I were a pupil of baccalaureate, I would ask my center that taught me chemistry well. Chemistry is related with two other sciences: biomedicine and materials science. Chemistry should be taught in a laboratory because it is an experimental science. Here we have the data, what conclusions can we take out from them? Here we have the conclusions, what data can we explain with them?

Application of engineered nanomaterials in environmental protection and the visionary future is presented in Chapter 10. The world of science and technology is moving towards a visionary era of scientific introspection and scientific vision. The challenges and the vision of environmental engineering science are immense and versatile. This treatise deeply comprehends with insight and deep scientific understanding of the truth of environmental remediation and unveils the areas of conventional and non-conventional environmental techniques. The author in this treatise reviews the success of environmental engineering science in degrading recalcitrant organic and inorganic compounds in industrial wastewater and drinking water. Heavy metal and arsenic groundwater contamination are a veritable scientific burden in many developing and developed nations around the world. Mankind's immense scientific and knowledge prowess, the provision of clean drinking water, and scientific excellence are all the torchbearers towards a new scientific order in environmental engineering science. Today nano-science and nanotechnology are integrated with diverse areas of science and engineering, which includes environmental engineering and water purification. The author in this chapter elucidates the various nanotechnology techniques in water/wastewater treatment. Heavy metal and arsenic groundwater contamination in developing and developed nations around the world are veritably destroying the scientific firmament of might and vision. Technology has few answers to these ever-growing concerns. In this chapter, the author also delineates the different environmental engineering techniques in tackling arsenic groundwater contamination. The other areas of research pursuit in this treatise are the application of nanotechnology and nanomaterials (NMs) in environmental protection. Scientific vision and scientific divination are evolving at a rapid pace in this century. These areas of modern science and modern engineering are elaborated in deep detail in Chapter 10.

The chemistry and sustainable development is discussed in Chapter 11. In Chapter 12, the ability of manganese ions and Mn-porphyrins as potential pro-oxidants were examined in ascorbate-induced hyaluronan (HA) degradation by means of rotational viscometry. Further, their effect was examined in HA degradation initiated by Cu(II) ions in the presence of ascorbic acid (AA) as a source of •OH radicals and alkoxy-/peroxy-type radicals. The addition of Mn(II) ions or Mn-porphyrins resulted in ascorbate-induced HA degradation. However, after the addition of Cu(II) ions to the HA solution with ascorbic acid and Mn(II) ions or Mn-porphyrins, the degradation of HA was promoted in a dose-dependent manner. HA degradation induced by Cu(II) ions and ascorbic acid in the presence of Mn(II) ions was a bit slower than in the presence of Mn-porphyrins. Production of •OH radicals in the latter mentioned HA oxidative system was detected by electron paramagnetic resonance (EPR).

Chapter 13 is dedicated to the consideration of obtaining processes and properties of magnetic copper carbon mesoscopic composites (Cu-C

MC), which contains phosphorus and then is modified with the addition of therapeutically active substances such as adenosine tri-phosphoric acid, ascorbic acid, and Urotropine. The producing of magnetic mesocomposites with additive bioactive substances is carried out on the following scheme: (1) the obtaining of copper carbon mesocomposite by mechanic chemical method owing to the joint grating of copper oxide powder with polyvinyl alcohol; (2) the modification of copper carbon mesocomposite by an analogous method with the ammonium polyphosphate (APP) using for the obtaining of phosphorus-containing copper carbon mesocomposite; (3) the connection of bioactive substances to the phosphorus-containing mesocomposite obtained. Phosphorus-containing copper carbon mesocomposite is a magnetic mesoparticle with a phosphorus oxide link for therapeutically active substances. Initial copper carbon mesocomposite consists of copper-containing clusters size of which is equaled to less than 25 nm and which is found in the carbon shell. The above shell contains the carbon fibers from polyacetylene and carbine fragments with unpaired electrons on joints of fragments. At the mechanic chemical interaction of copper carbon mesocomposite (MC) with ammonium polyphosphate (APP), the phosphorus oxidation state change as well as the change of magnetic characteristics for the copper cluster are observed.

The maximum atomic magnetic moment of copper is equaled to 4.5 μB and its obtained at the relation MC to APp. correspondent to 1:0.5. In this case, the quantity of unpaired electrons on the carbon shell of the nanosized granule is increased in ten times. This fact can testify about the possibility of free radical activity decreased in the striking parts of the organism. The connection of the above said bioactive substances takes place owing to the active carbon shell by means of the mechanic chemical method with the small energetic expenses. At the same time, the high magnetic characteristics are preserved at the unpaired electrons presence on the carbon shell. Therefore, the obtained magnetic mesoscopic composites with therapeutically active substances can be considered as transport for bioactive substances and also as the inhibitors of radical processes in organisms.

Science, pseudo-science, and chronobiology are discussed in Chapter 14. In Chapter 15, pre-treatment of rusted steel surfaces with phosphoric acid solutions investigated in detail. And finally, the concepts, approaches, and techniques involved in the determination of the transport properties (TP) in solutions (diffusion, conductivity, and viscosity), their importance in the scientific and technologic communities, and the critical analysis of these data are focused in Chapter 16.

CHAPTER 1

Electrospinning of Bio-Based Polymeric Nanofibers for Biomedical and Healthcare Applications

RINKY GHOSH, [1] VEERESHGOUDA SHEKHARAGOUDA NARAGUND,[2] and NEHA KANWAR RAWAT[3, 4]

[1]*Department of Materials Science and Engineering, Indian Institute of Technology Kanpur, Kanpur – 208016, India*

[2]*Materials Science Division, CSIR-National Aerospace Laboratories, Bangalore, India*

[3]*Senior Lecturer, Chemistry, Haryana Education Services, Haryana, India*

[4]*Materials Research Laboratory, Department of Chemistry, Jamia Millia Islamia, New Delhi, India, E-mail: neharawatjmi@gmail.com*

ABSTRACT

The interest in the electrospinning process for the medicinal chemistry and nanotechnology field has escalated recently. This is because of simplicity in its design, cost-effectiveness, and interconnected porosity, better encapsulation of therapeutic agents, thermal consistency, higher loading capacity, and controlled release of drugs/gene carriers. Electrospun scaffolds recently have high behest for successful fabrications of scaffolds due to its significant endocytosis interactions with the cells and suitable degradation profile, tremendously exposed in the implants surgery and tissue regenerations.

Scaffolds of biocompatible nanofibers offer an efficient sustainable route for easy production of and facilitate the tailoring of surface properties by grafting and functionalization with bioactive agents. The biological studies have shown high potential in applications of biocompatible and biodegradable nanofiber scaffolds for wound dressing patches and drug delivery.

This chapter briefly discusses the electrospinning fundamental approaches to produced nanofibers scaffolds, biocompatible properties, drug-releasing rate kinetics, and its various applications in the interdisciplinary field of medicinal science.

1.1 INTRODUCTION

Innovative and scientific technologies focused around bio-based materials have gained considerable attention recently. They are currently at a peak of exigency to decline the usage of petroleum products and to raise the demands for bioplastics. The petrochemical products are now an integral and substantial part of society. The fossil fuel-based products can also be termed as never-ending life cycle [1] products causing a devastating effect to the natural habitat leading to climate changes, which create an environment pressure are now escalating day-by-day and the crude oil reserve prevalent currently coming to an end. The overall growth in the petrochemicals products is underpinning due to difficulty in finding an alternative for various structural applications. With consistent rise in demand of plastics on the per capita basis has lead to some major consequences around the globe creating permanent environmental pollution which needs to be addressed immediately. This has forced governments to curb the single-use plastics and focus more value chain of production activities to minimized effluents, wastage, and emissions rate during conversion of crude oil to the desired plastics products. The need for imposing a strict rule for waste management and recycling process is an urgent need that will not only maintain ecological balance but also helps in sustainable growth [1]. Today bio-based polymers are of pivotal importance and a popular choice for number of applications due to its sustainable benefits. High-performance properties of the bioplastics can be achieved by significant use of the nanofibers or biofibers [2] as composite materials or filler particles or as an additive. Before the starting of this new era, nanofibers are of vital importance and have gained growing scientific recognition, technological importance in the field of aviation, automobile, biomedical, etc. Electrospinning has been successfully used in the past few decades to produce nanofibers ranging from few micrometers to nanometers. It provides an opportunity for blending, incorporation of nanofillers into electrospun nanofibers; tune the surface properties and making fabrication more economically feasible and cheaper.

1.2 HISTORICAL EVIDENCE

It was unfamiliar for most of the researchers around the world during the beginning of this new era that electrospinning can enable to us produce nanofibers in the form of continuous filaments with diameter ranges from submicron to nanometers. The word 'electrospinning' has been derived from the term 'electrostatic spinning' and this process was first patented by Anton Formhals in the year 1934. He disclosed complete setup to produce polymeric filaments from the polymeric solution of cellulose acetate using a phenomenon of opposite polarity effects that generates an electrostatic effect.

1.3 A BRIEF ABOUT ELECTROSPINNING FUNDAMENTALS AND ITS VARIOUS PARAMETERS

The electrospinning set up mainly consists of two electrodes: one dipped into the polymeric solution and the other connected to the collector/rotatory plate. With the application of high voltage ranging from 10–25 kV to the polymeric solution, the electrical charges flow through the spinneret to the tip of the capillary tube. As the electric current encounters the solution at the tip end, this induced separation of charges on the surface of the droplet leads to the deformation of the hemispherical surface to the conical shaped (Taylor Cone). Once it exceeded the cohesive force or surface tension of the liquid, a thin jet emerges out and subdivides collecting on the moving rotatory drum which is kept at some suitable distance. Based on the reported articles, spinneret has been considered as an essential part of nanofibers production. Nanofibers with various diameters and multiple configurations can be obtained by slight variations in the processing parameters and core-shell implementation of the spinneret part. The salient features of electrospun nanofibers are dependent on various parameters-viscosity, voltage, surface tension, flow-rate, tip-to-collector (TCD) distance, temperature, and relative humidity. These essential parameters have a major impact on the diameter and morphology as well as the mechanical properties of the produced nanofibers. Through carefully conducted experiments and optimizations of the experimental set up allow hindering the possible contamination of produced nanofibers useful in wound healing treatment and its effective interconnected porous feature promotes cell adhesion-vital important phenomenon to improve biological functions and improved wound healing rate [12–16]. Figure 1.1 illustrates the application of electrospun nanofibers in various fields.

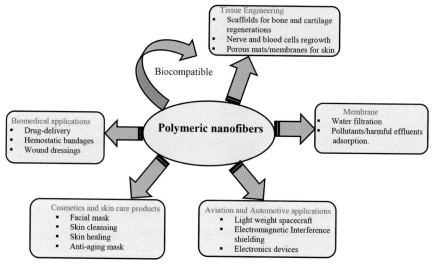

FIGURE 1.1 Various properties and applications of electrospun polymeric nanofibers.

The unoriented fibrous structure and high SVR mimicking the morphological 3D characteristics of the skin's extracellular matrix [17, 18] is useful for muscles and nervous tissue regeneration and nourishes wound segments thereby provide a moist environment for collagen resynthesis and enhancement of angiogenesis.

1.3.1 ELECTROSPINNING PROCESS

A schematic of an electrospinning setup is shown in Figure 1.2. In an electrospinning process, a high voltage electric field (EF) is applied between a polymer solution and a metallic collector. When the voltage crosses a limiting value, the charges in the polymer solution overcome the surface tension forces and a jet is ejected. The jet due to its hydrodynamic instabilities subdivides into millions of nanofibers as it travels towards the collector. The solvent evaporates quickly due to the large surface area of nanofibers and fibers are collected on the collector. With increasing time, the membrane thickness increases and can be peeled off from the collector.

The parameters affecting the electrospinning process are of great interest because, with the understanding of these parameters, it is possible obtaining nanofibers with different morphology and diameters. It is also possible to

Electrospinning of Bio-Based Polymeric Nanofibers

optimize setups to yield fibrous structures of various forms and arrangements. If one observes the electrospinning the three things that stand apart are: (i) the solution used, (ii) the capability of the electrospinning equipment, and (iii) the environmental conditions at which electrospinning is occurring. Thus, the parameters affecting the process can be categorized into two classes:

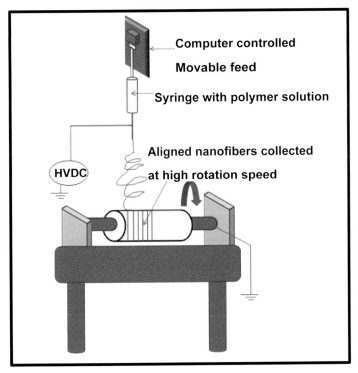

FIGURE 1.2 A schematic diagram of electrospinning equipment using a rotating drum collector to obtain aligned nanofibers.

1. The material system parameters:
   i. The polymer parameter:
      - The molecular weight of the polymer granules;
      - The weight distribution and architecture of the polymer (branched, linear, etc.).
   ii. The solution parameters:
      - Viscosity: High enough to maintain continuity within fibers and low enough to come through the capillary;

- The conductivity of solution;
- The dielectric constant of the solvent;
- Surface tension;
- The charge carried by the spinning jet.

2. The processing parameters: These pertain to the electrospinning processing conditions that can be controlled by the operators. Some electrospinning equipment support precise control over environments within the enclosure:

   i. The electric potential applied;
   ii. The flow-rate of solution;
   iii. Distance between capillary tips to a collector;
   iv. Ambient parameters (temperature and pressure);
   v. The motion of the target screen.

Therefore, in electrospinning, the above conditions can be varied to produce nanofibers with different morphologies/diameters and then optimize the parameters to produce highly aligned nanofibers.

1.3.2 METHODS TO OBTAIN ALIGNED NANOFIBERS VIA ELECTROSPINNING

Aligned nanofibers are obtained by many researchers using different techniques applying modifications to the conventional electrospinning equipment. These can be classified into three categories: (i) mechanical, (ii) electrical, and (iii) magnetic. All of these may involve one or more of (i) modified nanofibers collector system, and (ii) use of additional electrodes to direct electrical or magnetic fields (MFs). Some of the techniques used are:

1. High speed rotating drum collector;
2. Disc type collector with sharp edge;
3. Wire drum collector;
4. Dual vertical collector;
5. Collection between flat magnetic bars/strips;
6. Secondary Wedge electrodes;
7. Post-processing operation such as stretching;
8. Patterned insulating gaps on quartz electrodes;
9. Use of auxiliary electrodes to manipulate EFs;

Electrospinning of Bio-Based Polymeric Nanofibers

10. Ring electrodes and EFs using the principle of right-hand thumb rule to align magnetic fibers;
11. Use of AC voltages.

Sarkar et al. [35] reported the use of combination of DC and AC potentials for minimizing the inherent fiber instability in the electrospinning process, calling it as "biased AC electrospinning," and obtained highly-aligned mats of polymer or composite polymer fibers. The applied AC voltage introduces alternating positively and negatively charged regions in the fiber, resulting in a decrease in electrostatic repulsion and an increase in fiber stability found that a very high degree of fiber uniformity and alignment could be achieved for each polymer/solvent system investigated in this study as long as the following two general process requirements were followed: (i) The magnitude of the DC bias must be less than half the total amplitude of the AC potential; (ii) The AC frequency should be between 500 and 1000 Hz for optimum fiber stability.

Matthews [36] noted that collagen fibrils electrospun onto a mandrel rotating at a rate of less than 500 rpm produced random fibers. On increasing the velocity of the mandrel to 4500 rpm (1.4 ms^{-1}), finally resulted in the deposition of fibrils along the axis of rotation.

Ashish Aphale et al. [38] investigated a new configuration of axial rotating disc collector setup that yielded a good amount of alignment in PVA nanofibers of around 200 nm in diameter embedded with CNT over the surface of rotating disc.

Another interesting report by Ajao et al. [37], they demonstrated the use of MFs to align PEO nanofibers. Their modified electrospinning setup consisted of cylindrical magnet covered with silicon wafers on all sides kept between the ground electrode and spinneret. The aligned nanofibers were only observed over the top portion of the silicon wafer.

Li et al. [39] obtained aligned nanofibers in insulating gaps of various shape such as triangular, rectangular, circular holes cut between quartz electrodes (Figure 1.3).

Katta et al. [43] fabricated Cu wire drum (d = 12.7 cm, l = 30 cm) rotating collector at 1 rpm to obtain macro alignment of 20% nylon nanofiber for shorter duration of times. The possible reason might be the charge deposition over the nanofibers. It is possibly due to an accumulation of charge in the fiber mat, but the jet was not observed to stray away from the grounded wires as is often the case with low conductivity polymers. It is also possible that the nylon mat acts as a conductor; the thickened mat on the wire drum fills in the gaps between the wires with a grounded surface and the electrospinning jet effectively experiences the grounded field of a solid drum.

FIGURE 1.3 Aligned nanofibers obtained by nanofibers obtained by dual vertical stainless-steel wire technique.

Source: Reprinted with permission from [43]. © 2004 American Chemical Society.

1.4 PROPERTIES AND VERSATILE BIOMEDICAL APPLICATIONS OF ELECTROSPUN BIODEGRADABLE NANOFIBERS

With the advent of nanotechnology and nanoscience, it has gained significant popularity and commercial interests with global productive benefits. Among various versatile techniques available scientists around the globe focuses more on electrospinning process to fabricate nanofibers, its remarkable properties have been intensively investigated due to its unique morphology and enhanced flexibility. Several publications of inexpensive electrospinning process for biomedical applications has grown exponentially due to its outstanding porous sizes or interconnected pore structure, ease of fabrication and large surface to volume ratio (SVR) facilitate the active loadings of biomolecules which helps in possible transportation of nutrients and other bioactive agents. The successful *in-vitro* and *in-vivo* studies suggested that biodegradable and biocompatible polymers for example Collagen, chitosan-chitin, poly-lactic acid (PLA), elastin, alginate, poly(ε-caprolactone), polyglycolide (PGA), fibrinogen, and polyesters, etc., have wide range of future aspects due to its significant endocytosis

Electrospinning of Bio-Based Polymeric Nanofibers

interactions with the cells and suitable degradation profile, tremendously acknowledged in the implants surgery and tissue regenerations. As they metabolize, at the same pace and also its degradation time are tunable with the tissue regrows time range. Literature, reported the overwhelming and garnering performance of electrospinning process although several others advanced technologies are available to fabricate nanofibers/nanofibrous mats some of them includes, template synthesis, drawing process, solvent-casting, stretching method, inland spinning and phase separation. But more importantly with the increased knowledge of nanostructures, nanoparticles, and nanofibers, highlighted research conclusions pointed-out in numerous articles suggested the issues regarding fabrications using other techniques but in spite of all electrospinning pave the way for advancements in many bioengineering areas due to its easy, cost-effective, and versatile processing techniques with scant drawbacks that can be tunable (Figure 1.4). Scaffold fabrications have been developed with the help of the electrospinning process as it also provides an efficient sustainable route for easy production of biocompatible scaffolds and its properties can be preferably tailored with suitable grafting or functionalization. Haider et al. have reported the intriguing performance and important characteristics of hybrid scaffolds successful in providing similar environment to the cells for better adhesion; attachment ultimately results in proliferation [30]. More efforts are still performed to widespread the electrospinning techniques and to scale up the process as it enables the processing of biodegradable polymeric solution of several liters in a single run, therefore has gained noteworthy interests and profitable business to be used for industrial applications.

From the biological viewpoint, human tissues and organs possessing hierarchical fibrous structures, therefore progressive researches focused more on the biopolymers based nanofibrous for bioengineering applications [4]. Electrospinning is the most reliable, versatile, ease of processing technology to develop highly efficient uniform surface area non-woven nanofibrous mats with small porous size are of physiochemical and biological importance competent in the field of tissue prosthesis, organ implantation, drug delivery, and wound dressings. To ideally design the scaffolds for tissue engineering applications is one of the greatest challenges, as such fibrous based matrix effectively fabricated that can imitate the native biological functions and provides a template for cell proliferation which helps in the regeneration of tissues/organs [4] (Figure 1.5). Therefore, it has been reported and surges in the number of repots depicting the effective utilization of biodegradable polymeric nanofibrous in the field of biomedical science by imparting a favorable cellular environment that can mimic the extracellular matrix [8] and provide a platform for the development of biomaterials scaffolds [5–7].

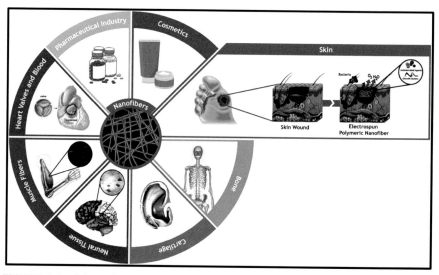

FIGURE 1.4 Schematic representation of various biomedical applications of electrospun biodegradable nanofibers.

Source: Reprinted with permission from [41]. © 2018 Elsevier.

The modern technology relies most on the fibrous hemostatic bandages for wound healing applications with desired pore size essential for the wound healing treatment [9]. It effectively declines the penetration of bacteria's and prevents the wound from its further attack because of its desirable pore diameter. With specific choice of biopolymers based fibrous bandages are new subject of interests to many researchers around the world as it provides several requisite promising properties in one single foundation such as antibacterial, antimicrobial, and it helps in remedying the pain by imparting anti-inflammatory conditions which enhance the rate of healing.

In the competitive market for cosmetics and skincare products polymeric nanofibers find greater recognition widely used for skin cleansing agent and facial mask having both therapeutic and medicinal effects [10]. As the nanofibers have very small interstices and pore diameter which are effective for rapid transfer of additives into the skin pores, penetrate painlessly and cover the topography of the skin-is well known for increasing the rate of skin healing and helps in the skincare treatment.

Electrospinning of Bio-Based Polymeric Nanofibers

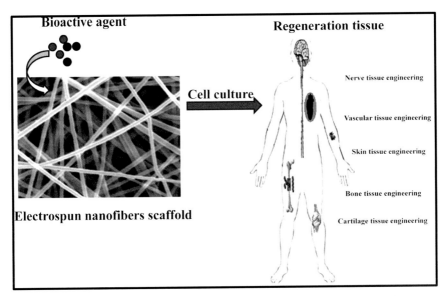

FIGURE 1.5 Schematic representation of tissue regeneration with the help of electrospun nanofibrous scaffolds.

Source: Reprinted with permission from [42]. © 2017 Elsevier.

The traditional and outrageous methods to stop bleeding and promote rapid hemostasis is by applying hand pressure or using chemical spray technique on the wounded topography that promotes vasoconstriction but change in clotting activity or secondary damaged to the wounds has been observed [11]. The clotting cloth available for rapid hemostasis have some drawbacks associated with it, for example, poor adhesion, less soothing effect, and a possible chance for bacteria's penetration.

In contradiction, electrospun nanofibrous mats developed is of crucial importance and have attracted much attention for the rapid hemostasis. They facilitate the blending of hemostasis components with polymeric solution resulting in mats formations have a large surface area and high porosity which promotes good adhesion. They inhibit the breeding of bacteria and further provide a favorable condition for the regrowth of cells by absorbing all the extracellular fluids [3], as shown in Figure 1.6.

Electrospinning of nanofibers involves the preparation of a suitable polymer solution using a particular compatible solvent of various viscosities. This technique of producing nanofibers increases the SVR ensures large porosity facilitates the growth of cells attachment and proliferation.

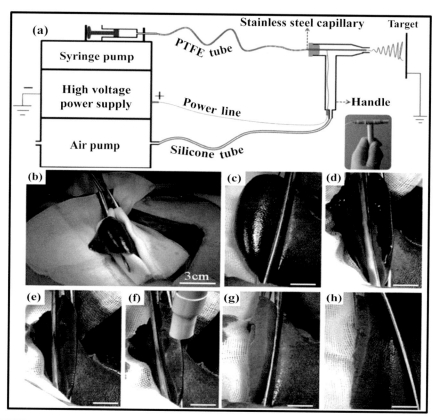

FIGURE 1.6 (a) Showing electrospinning technique to produce nanofibrous mats. (b) to (h) Schematic diagram of rapid hemostasis and rapid healing effect using nanofibrous mats.
Source: Reprinted with permission from [32]. © 2017 Elsevier.

1.5 HEALTHCARE APPLICATIONS OF BIO-BASED POLYMERIC ELECTROSPUN NANOFIBERS

1.5.1 TISSUE ENGINEERING/SKIN REGENERATION AND WOUND HEALING

Since a decade, the main tissue engineering application lies in the healing of physiological disorders which have a devastating effect on the nervous, bone, and spinal tissues, etc., to seek repair and replace tissues by appropriate incorporation of biologically active molecules. Transplantations of organ/tissue are a major health issue as firstly it is not economical, secondly limited accessibility of donor matching. Skin burns/wounds have its own fundamental aspects to self-heals or having significant self-regenerating ability by maintaining the bodies homeostasis conditions unless the consequence of chronic wounds is severe or unpleasantly extensive lesions [19]. Since many of the effective treatments are currently available sometimes failed to treat the wounds completely causing trans epidermal water loss and microbial growth and as a result infection spreads swiftly have a major impact which leads to patient's death. The other important factor to be considered is bio-degradability, as old tissue scaffolds must be replaced with new ones with a passage of time. Thus, to make them skilled in expediting this novel cell-biomaterial interfaces. Scaffolds are the most essential part and are highly suitable for tissue regeneration as the morphological structure contains microscale interconnected pores topographically similar features to that of ECM presents in our body which possessed 3D porous structure and exclusively made up of collagen and others biocomponents. Several authors have recently reported the incorporations of proteins to the interconnected pores of the electrospun nanofibrous mats. Proteins are known to be produced crucial biological signals beneficial for tissue regenerations particularly helpful in the regrowth of nervous tissue. PCL nanofibers are amenable for incorporation of protein molecules due to its large topographical network are promisingly attractive with improved depositions efficiency because of its physiological, biocompatible features and degradation rate which is closely matched with the rate of regeneration of new tissues obliging for scaffolds preparation and drug-delivery applications. The fundamental goal is to develop neotissues/neoorgans with similar biomimetic geometries as native one. One drawback of using PCL is its poor hydrophilic nature leads to reduction of cell migration, cell attachment, proliferation, and differentiation [22, 23]. The need for arterials prostheses is now in urgent demand due to failure of vascular system its main cause lies in cardiovascular disease which leads to stroke, coronary, inflammatory, and infraction. The blood vessels are primary source of transfer of nutrients and maintaining temperature in human body. Fabrication of vascular tissues is the most fundamental challenges faces through the decade, but with advent growth of electrospinning process and use of composite biodegradable polymers have reduced the chances of deaths. Valence et al. has successfully reported the electrospun scaffolds composed of PCL and cross-linked with acrylate L-lactide-co-trimethylene carbonated elastomer depicted burst pressure, suture retention strength and negligible formation of aneurysm show striking resemblance with human veins confirmed through *in-vivo* studies [24, 25].

Jayakumar et al. were reported the electrospun fibrous materials-chitosan/chitin based used for posttraumatic wound healing applications, tissue engineering scaffolds, and drug delivery [20, 21]. Chitosan/chitin possessed certain properties such as biocompatibility, low cytotoxicity, high durability, microfluids absorption, biodegradability, and antimicrobial-antifungal activity, causing an accelerating effect on the wound healing rate. But major drawbacks associated with it using neat chitosan lies in its poor mechanical properties and loss of integrity in aqueous media restricted their applications in biomedical fields. To overcome these disadvantages of single-use electrospun polymer, researchers across the globe are more interested in the composite scaffolds to avoid fishnet effects and to promote cell proliferation and cell adhesion.

Fibrous nanocomposites (NCs) composed of hydroxyapatite (HA) and silk fibroin (SF) owing to its flexibility, low inflammatory responses, biocompatibility, and show better permeability towards oxygen and water vapor. Thus emulating the basic building blocks structure of collagen nanofibers present in the orthopedic area specifically used as a substitute for bone replacement and dental implantation due to its weak osteogenic capacity. SF is a fibrous protein obtained from the silk-worm in the filamentary form. Bone formation required the requisite mesenchymal stem cells derived from bone marrow and was a cell cultured in osteogenic media for 30 days at ambient conditions. The novel biomimetic SF composite scaffolds prepared by electrospinning, aid the growth of mesenchymal stem cells demonstrated the significant bone formation as shown in Figure 1.7.

1.5.2 WOUND DRESSING

Various biodegradable natural and synthetic polymers electrospun into nanofibers facilitates the cell proliferation of fibroblasts and cell attachment with the additional benefit of tunable mechanical properties which prevent the wound contraction during implantation and has significantly accelerated the fabrication of scaffolds for wound healing applications.

Kumber et al. have reported the efficacy for dermal regrowth, improvement in cell proliferation and effective propagation of human skin fibroblasts through scaffold fabrication using polylactic acid (PLA) and poly lactic-co-glycolic acid (PLGA) [26]. The entire outcome of the process was investigated through *in-vivo* studies and has demonstrated the biodegradable scaffolds made of PLA/PLGA are best till date for wound healing treatment, as it supported the growth of keratinocyte, fibroblast, and endothelial cell. The *in-vivo* studies revealed that the slight variations in the mole fractions of nanofibers of lactide/glycolide (85:15, 75:25, and 50:50) would significantly result in better up-gradation in the expression of collagen and fibroblasts growth [27]. He has shown the mole fraction ratio of nanofibers 85:15 and 75:25 remained stable after 1 year of implantation and therefore considered it as a favorable scaffold for wound treatment whose degradation rate closely matches with the healing rate in defected tissue as shown in Figure 1.8.

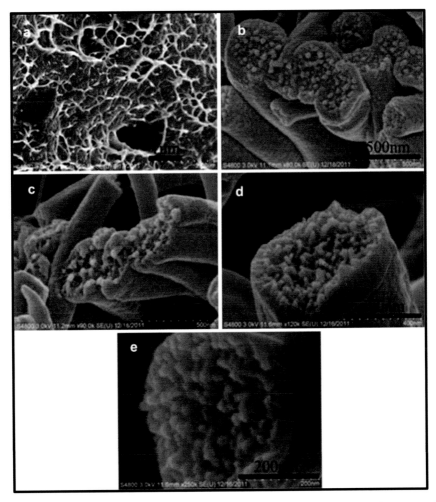

FIGURE 1.7 (a–e) Represents the SEM micrograph of nanocomposite scaffolds composed of Silk fibroin (SF) and hydroxyapatite particles which are well exposed to the surface of the scaffolds.

Source: Reprinted with permission from [33]. © 2012 Elsevier.

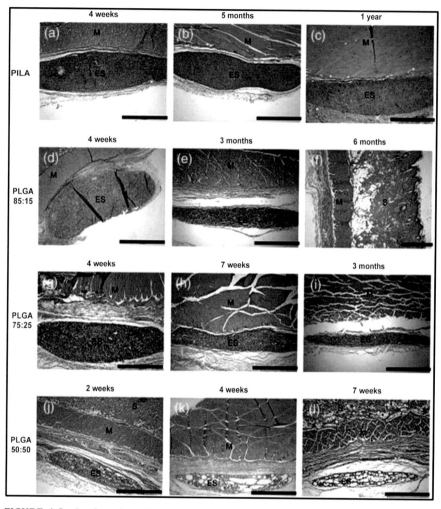

FIGURE 1.8 Implanted scaffolds composed of electrospun PLA and PLGA of varying molar ratios into the Wistar rats and the tests conducted from 2 weeks to 2 years to see the fibroblasts growth.

Source: Reprinted with permission from [34]. © 2015 Elsevier.

1.5.3 DRUG DELIVERY

Many therapeutic compounds are incorporated into synthetic or natural bio-based electrospun scaffolds as a drug carrier for preferable delivery of antibiotics and anticancer drugs in the preferred sections of our body. Electrospun nanofibers have shown inevitably controlled at the release of drug at the definite period of time and hence a suitable prospect for drug delivery applications. The rate of release of drug profile can be tunable by controlled modulation of morphology, interconnected pore size, short diffusion length, and drug-binding sites which ensure better therapeutic effects. The various attractive features of electrospinning nanofibers such as mechanical modulus, hydrophilicity, higher drug loading capacity, higher encapsulation efficiency, cost-effectiveness, and drug-releasing kinetics can easily be determined by the drug diffusion analysis and weightage average of distribution of drug particles on the surface of the nanofibers. Nanofibers scaffolds used as a tool for drug delivery and postoperative chemotherapy due to its excellent barrier properties and biocompatibility.

To design the proper drug release system, blending of drugs into polymeric solutions is a primary requirement before performing the electrospinning process to obtained ultrafine drug-loaded scaffolds. The drug loaded-scaffolds exhibited burst release at the initial stage of the installations to eliminating the effect of bacterial attacks before proliferation [28].

Huang et al. have successfully demonstrated the importance of Gelatin nanofibers for drug delivery applications [29]. Gelatin is a natural-based biodegradable polymer derived from collagens a potential candidate for controlled release of drugs has several advantages compared to conventional pharmaceutical dosage materials available in the market such as desired therapeutic effect, low cytotoxicity, enhanced biocompatibility, etc.

Yuan et al. have demonstrated a pH-sensitive polylactide nanofibers comprised of sodium hydrogen bicarbonate ($NaHCO_3$) used for controlled release of drug carrier ibuprofen, resulted in faster rate of skin wound healing. The pH-sensitive stimuli-responsive behavior of electrospun nanofibers are well justified and shown rate kinetics of drugs can be enhanced by using smart electrospun nanofibers in terms of drug release [31] (Figure 1.9).

1.6 CONCLUSION

Electrospinning is the most versatile technique for making of continuous nanofibers and showed great potential in the emerging field of biomedical applications. With the advent of new technological advancement to produce ultrafine nanofibers, electrospinning has manifested new route for exploring both synthetic and natural biodegradable polymers for the fabrications of peerless scaffolds. Tremendous researches are going on this ancient known technology to developed nanofibrous scaffolds and with proper selection of polymeric materials as its play key role in the fabrication process broadly applied as the therapeutic agents and tissue regenerations. The more reliable, realistic, and judiciary pathway facilitates the incorporation of drug loadings and mimics the native ECM due to microstructures and morphological similarities and promisingly useful for future biomedical applications such as tissue engineering, asymmetric dressing and controlled release of drug carriers (genes). Furthermore, effective modifications of scaffolds with the functional agents fine-tune the surface properties of scaffolds offer the versatility and unique nanostructures potential for biomedical applications.

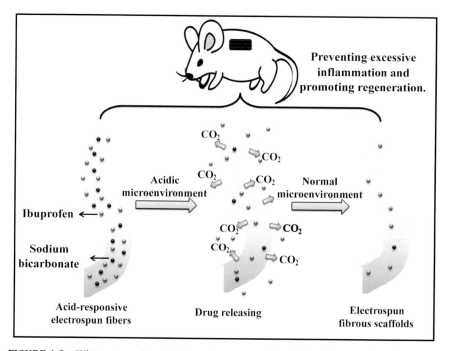

FIGURE 1.9 Wistar rat model and schematic representation of stimuli-responsive electrospun nanofibers.

Source: Reprinted with permission from [40]. © 2014 Royal Society of Chemistry.

However, despite many achievements and numerous studies which have successfully demonstrated the promising characteristics of electrospun nanofibers *in-vitro* studies, but few articles reported the *in-vivo* studies of nanofibers scaffolds which reveals the further need for nanofibers meshes for drug delivery and healing applications. In other words, this exciting research area needs new avenues and further requirement of more complicated and sophisticated design to meet the demands for diagnostic and therapeutic applications.

KEYWORDS

- **biomedical applications**
- **drug-delivery applications**
- **electrospinning**
- **grafting**
- **nanofibers**
- **nanotechnology**
- **scaffolds**
- **sustainable route**
- **therapeutic agents**
- **tissue regenerations**
- **wound dressings**

REFERENCES

1. Lucia, L. A., Argyropoulos, D. S., Adamopoulos, L., & Gaspar, A. R., (2007). Chemicals, materials, and energy from biomass: A review. *ACS Symp. Ser., 954*, 2–30.
2. Ramamoorthy, S. K., Skrifvars, M., & Persson, A., (2015). A review of natural fibers used in biocomposites: Plant, animal and regenerated cellulose fibers. *Polym. Rev., 55*, 107–162.
3. Bao, J., Yang, B., Sun, Y., Zu, Y., & Deng, Y., (2013). *J. Biomed. Nanotechnol., 9*, 1173–1180.
4. Huang, Z. M., Zhang, Y. Z., Kotaki, M., & Ramakrishna, S., (2003). A review on polymer nanofibers by electrospinning and their applications in nanocomposites. *Compos. Sci. Technol., 63*, 2223–2253.
5. Buchko, C. J., Chen, L. C., Shen, Y., & Martin, D. C., (1999). Processing and microstructural characterization of porous biocompatible protein polymer thin films. *Polymer, 40*, 7397–7407.
6. Fertala, A., Han, W. B., & Ko, F. K., (2001). Mapping critical sites in collagen II for rational design of gene-engineered proteins for cell-supporting materials. *J. Biomed. Mater Res., 57*, 48–58.

7. Yusop, S. M., Nadalian, M., Babji, A. S., Mustapha, W. A. W., Forghani, B., & Azman, M. A., (2016). Production of antihypertensive elastin peptides from waste poultry skin. *ETP Int. J. Food Eng., 2,* 21–25.

8. Boudriot, U., et al., (2006). Electro spinning approaches toward scaffold engineering: A brief overview. *Artif. Organs, 30,* 785–792.

9. Zahedi, P., Rezaeian, I., Ranaei-Siadat, S. O., Jafari, S. H., & Supaphol, P., (2010). A review on wound dressings with an emphasis on electrospun nanofibrous polymeric bandages. *Polym. Adv. Technol., 21,* 77–95.

10. Huang, Z. M., Zhang, Y. Z., Kotaki, M., & Ramakrishna, S., (2003). A review on polymer nanofibers by electrospinning and their applications in nanocomposites. *Compos. Sci. Technol., 63,* 2223–2253.

11. Liu, M., Duan, X. P., Li, Y. M., Yang, D. P., & Long, Y. Z., (2017). Electrospun nanofibers for wound healing. *Mater. Sci. Eng. C., 76,* 1413–1423.

12. Braghirolli, D. I., Steffens, D., et al., (2014). *Drug Discovery Today, 19,* 743–753.

13. Bhardwaj, N., & Kundu, S. C., (2010). *Biotechnol. Adv., 28,* 325–347.

14. Huang, Z. M., Zhang, Y. Z., et al., (2003). *Compos. Sci. Technol., 63,* 2223–2253.

15. Zhong, S., Zhang, Y., et al., (2010). *Wiley Interdiscip. Rev. Nanomed. Nanobiotechnol., 2,* 510–525.

16. Kumbar, S., James, R., et al., (2008). *Biomed. Mater., 3,* 034002.

17. Vasita, R., & Katti, D. S., (2006). *Int. J. Nanomed., 1,* 15.

18. Pham, Q. P., Sharma, U., et al., (2006). *Tissue Eng., 12,* 1197–1211.

19. Pereira, R. F., & Bartolo, P. J., (2016). *Adv. Wound Care, 5,* 208–229.

20. Jayakumar, R., Menon, D., Manzoor, K., Nair, S. V., & Tamura, H., (2010). Biomedical applications of chitin and chitosan based nanomaterials: A short review. *Carbohydr. Polym., 82,* 227–232.

21. Fuller, B., (2019). Role of PGE-2 and other inflammatory mediators in skin aging and their inhibition by topical natural anti-inflammatories. *Cosmetics,* 6.

22. Chen, H., Peng, Y., Wu, S., & Tan, L. P., (2016). Electrospun 3D fibrous scaffolds for chronic wound repair. *Materials (Basel), 9,* 1–12.

23. Li, W. J., Cooper, J. A., Mauck, R. L., & Tuan, R. S., (2006). *Acta Biomater., 2,* 377–385.

24. De Valence, S., Tille, J. C., Mugnai, D., Mrowczynski, W., Gurny, R., Möller, M., & Walpoth, B. H., (2012). *Biomaterials, 33,* 38–47.

25. Stefani, I., & Cooper-White, J. J., (2016). *Acta Biomater., 36,* 231–240.

26. Kumbar, S. G., Nukavarapu, S. P., James, R., Nair, L. S., & Laurencin, C. T., (2008). Electrospun poly(lactic acid-co-glycolic acid) scaffolds for skin tissue engineering. *Biomaterials, 29,* 4100–4107.

27. Zuber, A., Borowczyk, J., Zimolag, E., Krok, M., Madeja, Z., Pamula, E., & Drukala, J., (2014). Poly(L-lactide-co-glycolide) thin films can act as autologous cell carriers for skin tissue engineering. *Cell. Mol. Biol. Lett., 19,* 297–314.

28. Cui, W., Zhou, Y., & Chang, J., (2010). Electrospun nanofibrous materials for tissue engineering and drug delivery. *Sci. Technol. Adv. Mater., 11,* 014108.

29. Huang, Z. M., Zhang, Y. Z., Ramakrishna, S., & Lim, C. T., (2004). Electro spinning and mechanical characterization of gelatin nanofibers. *Polymer, 45*(15), 5361–5368.

30. Haider, A., Haider, S., & Kang, I. K., (2018). A comprehensive review summarizing the effect of electrospinning parameters and potential applications of nanofibers in biomedical and biotechnology. *Arab. J. Chem., 11,* 1165–1188.

Electrospinning of Bio-Based Polymeric Nanofibers

31. Yuan, Z., Zhao, J., Chen, Y., Yang, Z., Cui, W., & Zheng, Q., (2014). Regulating inflammation using acid-responsive electrospun fibrous scaffolds for skin scar less healing. *Mediators of Inflammation*, 858045.

32. Liu, M., Duan, X. P., Li, Y. M., Yang, D. P., & Long, Y. Z., (2017). Electrospun nanofibers for wound healing. *Mater. Sci. Eng. C., 76*, 1413–1423.

33. Ming, J., & Zuo, B., (2012). A novel electrospun silk fibroin/hydroxyapatite hybrid nanofibers. *Mater. Chem. Phys., 137*, 421–427.

34. Kai, D., Liow, S. S., & Loh, X. J., (2015). Biodegradable polymers for electrospinning: Towards biomedical applications. *Mater. Sci. Eng. C., 45*, 659–670.

35. Sarkar, S., Deevi, S., & Tepper, G., (2007). Biased AC electrospinning of aligned polymer nanofibers. *Macromol. Rapid Commun., 28*, 1034–1039.

36. Matthews, J. A., Wnek, G. E., Simpson, D. G., & Bowlin, G. L., (2002). Electrospinning of collagen nanofibers. *Biomacromolecules, 3*, 232–238.

37. Ajao, J. A., Abiona, A. A., Chigome, S., Fasasi, A. Y., Osinkolu, G. A., & Maaza, M., (2010). Electric-magnetic field-induced aligned electrospun poly (ethylene oxide) (PEO) nanofibers. *J. Mater. Sci., 45*, 2324–2329.

38. Aphale, A., Macwan, I., Bhosale, S., Patel, S., Kuram, S., Venugopal, V., & Patra, P. (2012). Fabrication of polyvinyl alcohol (PVA) and CNT filled PVA *Nanofibers by Electro Hydro Jets Spinning.*

39. Li, D., Ouyang, G., McCann, J. T., & Xia, Y., (2005). Collecting electrospun nanofibers with patterned electrodes. *Nano Lett., 5*, 913–916.

40. Yuan, Z., Zhao, J., Zhu, W., Yang, Z., Li, B., Yang, H., Zheng, Q., & Cui, W., (2014). Ibuprofen-loaded electrospun fibrous scaffold doped with sodium bicarbonate for responsively inhibiting inflammation and promoting muscle wound healing *in-vivo*. *Biomater. Sci., 2*, 502–511.

41. Miguel, S. P., Figueira, D. R., Simões, D., Ribeiro, M. P., Coutinho, P., Ferreira, P., & Correia, I. J., (2018). Electrospun polymeric nanofibres as wound dressings: A review. *Colloids Surfaces B Biointerfaces., 169*, 60–71.

42. Liu, M., Duan, X. P., Li, Y. M., Yang, D. P., & Long, Y. Z., (2017). Electrospun nanofibers for wound healing. *Mater. Sci. Eng. C., 76*, 1413–1423.

43. Katta, P., Alessandro, M., Ramsier, R. D., & Chase, G. G., (2004). Continuous electrospinning of aligned polymer nanofibers onto a wire drum collector. *Nano Lett., 4*, 2215–2218.

CHAPTER 2

Solar Energy for Green Engineering Using Multicomponent Absorption Power Cycle

SATCHIDANAND R. SATPUTE, [1] NILESH A. MALI, [2] and
SUNIL S. BHAGWAT[3]

[1]*Department of Chemical Engineering, Vishwakarma Institute of Technology, Bibwewadi, Pune, India, E-mail: satchidanand.satpute@vit.edu*

[2]*Chemical Engineering Division, National Chemical Laboratory, Pashan, Pune, India*

[3]*Department of Chemical Engineering, Institute of Chemical Technology, Matunga, Mumbai, India*

ABSTRACT

Energy is a bottleneck for the development of modern civilization. Green energy becomes a way towards the sustainable future of human society. The increasing demand of energy and scarcity of the energy resources is directing us towards the innovative ways for the generation of power. Solar energy is abundantly available; hence, harnessing solar energy efficiently would be most preferable option for power generation. This thought has given rise to new methodologies to study the thermodynamic systems. Established power generation uses steam power plant where coal is used primarily as heat source works in range of temperature 540–46°C and pressure of 150 bar–0.1 bar. Using solar easily steam can be produced up to 150°C. This low temperature energy available couldn't be effectively used using traditional steam power plants.

Majority of power plants worldwide produce power using the Rankine cycle that are driven by various kinds of non-renewable heat sources (viz. coal, oil, gas, etc.) using steam as working fluid. However, use of mixture as working fluid instead of the single fluid could improve thermal efficiency of the cycle [1, 2]. The mismatch of temperatures of contacting fluids in regular steam power plant causes irreversibility which can be reduced by using the multi-component working fluid [3, 4]. The boiling and the condensation of binary fluid occurs over variable temperature range, compared to single temperature for single fluid and can be understood clearly with Lorentz efficiency comparison with Carnot efficiency [5, 6]. Using multi-component working fluid with multi-temperature boiling, one can reduce the heat transfer related irreversibilities and therefore improve the heat source utilization. Malony and Robertson (1996) were the first to conceive the idea to use ammonia water mixture as working fluid for power generation. Kalina [7] first conceived the idea of using ammonia water as working fluid for the usage of heat for the gas turbine in a combined cycle power generation system. EI-Sayed and Tribus [8] found that Kalina cycle can have 10–30% higher thermal efficiency than equivalent Rankine cycle. Ibrahim (1992) and Sonntag (1991) have also analyzed the Kalina cycle. Their studies have showed the advantage of the Kalina cycle over the conventional Rankine cycle under certain conditions.

This chapter deals with different efforts starting with traditional ways of producing power. The efficiencies pros and cons of various methods for power generation will be compared based on 1st law, 2nd law efficiencies, exergy efficiency, etc. The cumulative analysis of power generation relating to the temperature, pressure of available heat source will be carried out. The most efficient way to harness low temperature energy using multicomponent mixture will be presented with elaborate analysis with unique platform of exergy relating to all different available methods and sources for power generation. The present work is patented with Bhagwat et al. [9].

2.1 INTRODUCTION

Economic growth of any nation directly depends on availability of energy at affordable prices. The environmental preservation is most important for sustainable growth. Hence, solar energy becomes the best green energy source for sustainable growth with clean environment. Solar energy is most abundantly used energy, available in Asian countries. The amount of energy released by sun in one day is sufficient for requirement of human civilization for a year. Solar energy is one of the most feasible green energy sources available.

Solar Energy for Green Engineering

2.1.1 SOLAR ENERGY AND GREEN ENGINEERING

Solar energy and green energy go hand in hand. The fossil fuel energy is traditional source of power requirement that has lead to environmental damage through global warming. For sustainable development, it is important to increase dependence on clean energy. Solar energy is best option for green sustainable energy because of it abundance, ease of use. Refrigeration and Power sectors are energy intensive fields. Use of solar energy in this field in terms of solar heat will be very much effective and beneficial. Power generation plants have maximum efficiency of approximately from 35% at power plant whereas there are additional 15% losses in transportation. Hence, effective efficiency of conversion of thermal to electricity is only 20% user end whereas direct thermal to power at user end will have minimum efficiency of 40%. This book chapter will normally focus on this concept of using solar heat locally to make use of solar energy much more efficiently than max 15% solar photovoltaic (PV) achievable efficiency.

2.1.2 NON-SOLAR-BASED ENERGY

Non-solar-based energy can be categorized majorly into renewable and non-renewable energy sources. Non-renewable sources mainly consist of fossils that include coal, petroleum. Whereas renewable energy (RE) sources include wind power, ocean thermal, biomass, etc. The non-renewable energy sources has advantage of high energy density and efficiency, continuous supply of energy whereas disadvantage of pollution causing health hazard and damaging the ecology to make it less unfit for living. The non-renewable energy sources come up with advantage of clean energy sources but disadvantage of low efficiency and non-continuity as available for limited time.

2.2 DIFFERENT WAYS OF HARNESSING ENERGY FOR POWER

2.2.1 THERMAL POWER PLANTS

Thermal power plants are traditional way of producing power using coal as heat source. Steam Rankine cycle is known to produce power in thermal power plant. The single component working cycle is termed generally (that is widely used) Steam Rankine Cycle system. Systems using steam as the working fluid and fossil or nuclear fuels as primary sources of energy are most widely employed for electric-power generation. The Rankine cycle usually employed in modern steam power-plant is much more complex and exhibits higher conversion efficiency than the original simple Rankine-cycle that was devised in the nineteenth century. Reheat and feed-water heating (i.e., regeneration) are the major modifications that have been introduced [10].

The Rankine cycle has been described in many engineering thermodynamics textbooks [11] it is briefly covered here. The simple Rankine cycle, shown in Figure 2.1, consists of four steps. The working fluid is pumped to a high pressure and circulated through the boiler. The fluid is boiled at a constant pressure in the boiler after which the high-pressure vapor produced is expanded through a turbine, thus extracting work from it. The vapor exiting the turbine is condensed in a condenser by rejecting heat to a cooling fluid. There are several modifications to the Rankine cycle that are used to achieve better efficiencies. These include superheating, reheating, and regeneration. Water (Steam) is the working fluid of choice for most vapor power cycles. Water works over a broad range of temperatures and pressures, has a large heat capacity, and is stable, safe, and very environmentally friendly. The energy sources used to generate steam include gas, coal, oil, and nuclear sources. A small percentage of steam power plants use geothermal and solar energy sources. Reheating of the steam before it enters the medium pressure (MP) turbine improves the performance of the Rankine cycle engine. The increase in efficiency is due to the increase in the average temperature at which the heat is added. However, the most significant benefit of using reheat is the drier steam obtained at the exhaust of the turbine.

Another modification of Rankine cycle which results in its improved efficiency is regeneration. In simple cycles, feedwater enters the boiler at a temperature significantly lower than the saturation temperature at the corresponding boiler pressure. Thus, the boiler needs to supply heat first to raise the feedwater temperature to its boiling temperature, and then heat is supplied for the actual vaporization. This initial heating of feedwater contributes to major irreversibility in the boiler as a large temperature gradient is present between the combusted product and feed water. This low-temperature heat addition reduces the thermal efficiency of the cycle [12].

This irreversibility is reduced by heating the feed water in a series of feedwater heaters by steam extracted from the turbines at selected stages. Although the steam bled from the turbines to heat feedwater reduces the power output from the turbines, there is an overall increase in the thermal efficiency of the power plant due to reduced fuel consumption in the boiler system.

Solar Energy for Green Engineering

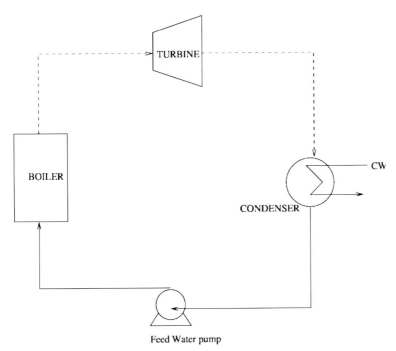

FIGURE 2.1 Simple Rankine cycle.

While steam is the working fluid of choice today, there are certain situations in which it does not work efficiently, particularly with low-temperature resources, e.g., case of a binary geothermal power plant. In such a plant, a fluid such as isobutane is boiled using a relatively low-temperature geothermal fluid, and used to spin a turbine working on a Rankine cycle. If steam were to be used in such applications (or if the plant is designed as a flash type plant), very low pressures (LPs) and large vacuums at turbine exit would result.

2.2.2 SUPERCRITICAL CYCLES

One of the present trends in the Steam Rankine cycle for improving efficiency is the use of supercritical cycles. The supercritical cycle includes high pressure and temperature exceeding critical point, i.e., pressure, and temperature of working fluid. Supercritical cycles have to operate at a higher pressure, since the boiler pressure has to exceed the critical pressure of the working fluid. Equipment costs go up at higher pressures, although there is an improvement in performance. The improvement in performance is because of increase of difference between sources and sinks temperature, pressures.

Supercritical cycles have been studied in the United States for geothermal applications as a part of the DOE Heat Cycle Research Program [13] and have been found to improve geofluid effectiveness (power output per unit mass of geothermal fluid consumed, usually expressed as kWh/lb), has also studied the use of supercritical operation in space applications. He suggested the use of a supercritical cycle in order to avoid the two-phase boiling encountered at subcritical conditions.

2.2.3 SOLAR THERMAL ENERGY

Solar thermal energy target solar energy utilization to produce steam and this steam can be utilized to produce power in Rankine cycle based power cycle. Temperature, pressure of steam achievable with solar thermal panel is important. There are four prominent technologies, i.e., parabolic troughs, central receiver tower, parabolic dish stirling engine system, linear Fresnel reflectors for steam power generation using solar energy. Parabolic trough has large parabolic troughs concentrating solar energy on central pipe carrying water. As heating proceed water gradually get heated to high temperature, pressure steam. It is simplest and cheapest way of producing steam using solar energy. Parabolic troughs have been reported to produce temperatures more than 260°C that is reasonable for production of power with Rankine cycle (Price Hank et al., 2002).

Central receiver tower for power generation could achieve temperatures approximately near 600°C which make it comparable with coal-based power plants. In the central receiver concept, huge numbers of solar panels are concentrated on the central tower. The solar concentrator can obtain temperature approximately 700°C and pressure of 200 bar. Central receiver tower concept claims thermal to the power efficiency of 50% using CO_2 based Brayton cycle with direct heating whereas using steam power plant could get the efficiency of 40% (Ho and Iverson, 2014).

2.2.4 SALT-BASED POWER PLANTS

Normally solar thermal power plant operates until the availability of sun in the day time. Option for utilizing solar energy 24×7 has given rise to the concept of salt-based power plants. The basic idea is storing solar energy with a salt-based solution that has high specific heat. Molten salt based on sodium, potassium are used as working fluid. Molten salt having very high thermal conductivity compared to water can store more heat with reduction in space. These molten salts are used to produce power with steam Rankine cycle (Vignaraboon et al., 2015).

2.2.5   ORGANIC RANKINE CYCLE

The low temperature heats sources are generally derive their energy from geothermal, solar or some waste heat sources. Solar energy can be used at various temperatures, depending on the collection method used. Low temperature heat sources have low availability of exergy and are not usually preferred as the energy sources. Such sources would be considered useful only if some economic advantage is found in their utilization. For instance, a geothermal power plant could prove to be economically feasible to supply an area in the vicinity of a geothermal steam field. Several power cycles that are suitable for use with low temperature heat sources have been proposed in the literature and have been used in practice. The Rankine cycle has been developed for use with low temperature heat sources by using low boiling working fluids such as organic fluids. Organic Rankine cycles (ORC) have been proposed and extensively used in geothermal power plants and low temperature solar power conversion.

Organic working fluids are found to be best choice for such applications of comparatively low temperature heat sources. Despite the fact that these fluids have lower heats of vaporization than water, which requires larger flow-rates, smaller turbine sizes are obtained due to the higher density at the turbine exit conditions. A variety of fluids, both pure components and binary and ternary mixtures, have been considered for use in ORCs. These include saturated hydrocarbons (HCs) such as propane, isobutane, pentane, hexane, and heptane; aromatics such as benzene and toluene; refrigerants such as R11, R113, R114; and some other synthetic compounds such as Dowtherm A. ORC have been proposed and used in a variety of application [14].

2.2.6   SOLAR THERMAL POWER

ORC plants can also be applied for the conversion of low and medium temperature (up to 300°C) solar heat. The Coolidge plant (Larson, 1997) that was a 200 kW plant built near the town of Coolidge, AZ, is a good example. For higher temperature heat sources, toluene seems to be a common choice of working fluid.

2.2.7 BOTTOMING CYCLE APPLICATIONS

ORCs have been proposed for the bottoming cycle in some applications. Binary mixtures have been found to have a better performance compared to pure fluid ORCs, [15, 16] on or Rankine cycle power plant.

2.2.8 BINARY POWER CYCLE USING AMMONIA WATER AS WORKING FLUID

Several cycles using ammonia water as working fluids for utilization of low temperature heat sources have been reported in literature. The common features of this cycle are boiling over range of temperature, than constant temperature boiling in Rankine cycles and condensation over a temperature range compared to constant temperature condensation in Rankine cycle. Few of this cycle are briefly discussed below.

2.2.8.1 MALONY AND ROBERTSON CYCLE

This is a simple cycle used for power generation using ammonia-water as the working fluid. It uses a single heat recovery unit in the cycle. As shown in Figure 2.2, after separation of vapor and liquid in the flash tank, vapors are superheated in the superheater and then expanded in the turbine isentropically. The vapors after expansion are absorbed with the weak liquid coming out of flash tank. The recuperator is only heat recovery device. It recovers the heat from stream leaving the flash tank. The heat source reported is at 167°C and heat sink is at 9°C. The thermal efficiency is reported as 50% of the Carnot efficiency achieved [17]. The Malony and Roberson's result has shown the Rankine steam cycle to be more efficient. This observation was because of limited range of ammonia water property data available to them.

2.2.8.2 KALINA CYCLE

This cycle operates as shown in Figure 2.3. Ammonia-water solution is heated and flashed in a tank. Vapors are condensed and pumped to boiler through heater. This mixture is then boiled completely to form vapors and it is then superheated in the superheater, followed by expansion in a turbine. Turbine exhaust gases (used for heat recovery by the recuperator) are taken to absorber where they are absorbed with the liquid returning from flash tank.

Solar Energy for Green Engineering

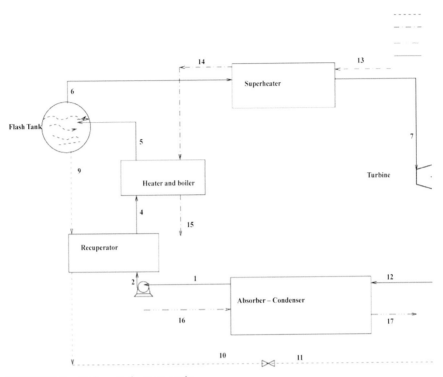

FIGURE 2.2 Malony Robertson cycle.

The binary mixture of ammonia-water is boiled at variable temperature in the boiler. The waste heat from turbine exhaust is used to preheat the fluid before entering the boiler. The high concentration (strong solution) and low concentration (weak solution) fluids are separated in the flash tank. The boiler gives the vapors at high pressure. This cycle considers expansion of vapors in turbine without condensation. The binary fluid gives better thermal match in reboiler resulting into less exergy loss. Efficient energy utilization allows for lower component cost of evaporator and condenser as area required reduces.

In Kalina cycle, the vapor composition entering the turbine is at mole fraction of 0.7. In present publication, the source temperature of 151°C is used and the heat is rejected at 15.6°C. Here addition of second condenser compared to the Malony Robertson cycle, adds an extra degree of freedom to the system. The thermal efficiency reported is of the order of 52% of Carnot efficiency. The vapor composition handled is of order of 0.7 (mole fraction) [18].

The one of the important features of Kalina cycle is extensive internal heat recovery and exchange arrangement that has resulted in minimization of irreversibilities in the heat transfer process and hence improvement of efficiency.

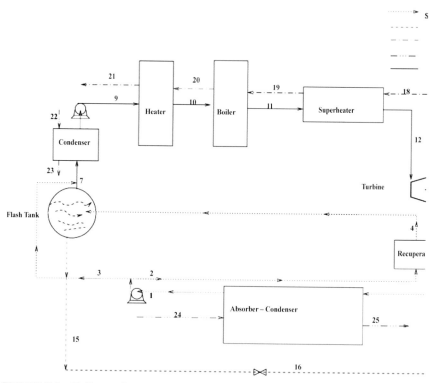

FIGURE 2.3 Kalina cycle.

The Kalina has also published series of patents with different aspect of Kalina cycle application with different range of temperature heat sources, few basic variations in his patents are discussed below.

2.2.8.2.1 Use of Variable Heat Source and Multiple Turbine for Power Generation

It separates rich stream and lean stream into two and heat recovery is done in the system and then they are recombined. It takes advantage of heating at variable temperature using variable temperature heat source as shown in Figure 2.4. This reduces irreversibility's in system.

Solar Energy for Green Engineering

Reboiler is used for heating rich liquid to high pressure and temperature. The rich vapors are finally separated and expanded in turbine. The weak liquor is heat exchanged and then used in the absorber to absorb the vapors from turbine after expansion. Heat source used here is at 522°C.

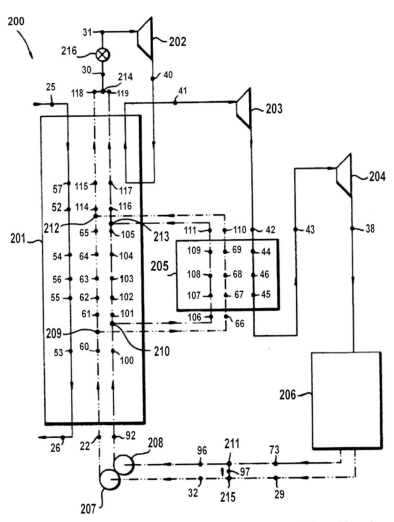

FIGURE 2.4 Use of variable temperature heat source and multiple turbines for power generation.

The multiple turbines are used with varying vapor composition and temperature at three different levels. The operating temperature level used for high pressure turbine is 497.55°C, 170 bar and is expanded to 357°C, 44 bar. The operating temperature for MP turbine is 497.55°C, 42.5 bar and is expanded to 306°C, 7.86 bar. The operating temperature for LP turbine is 439°C, 7.72 bar and is expanded to 86.67°C, 2.33 bar. This cycle has reported efficiency of 67% of Carnot efficiency.

In all patents mentioned in here, turbine operation is reported with temperature order of 500°C, pressure of order of 150 bar to expanded to lowest pressure of 3 bar and temperature of 70°C. The efficiencies achievable are of order of 65% of Carnot efficiency (Patent 509708, 4899545, 4586340, 4763480, 5822990).

2.2.8.2.2   Use of Variable Low Temperature Heat Source

This patent reports the use low temperature heat source and operation at low temperature and pressure level compared to other published patents as shown in Figure 2.5. It uses heat source at 168°C, going down to 56°C. It uses distillation column to get high composition liquid product. The distillation column is operated at LP. The liquid enriched product is pumped to evaporator and vapors are generated at desired temperature of 153°C and 20 bar. In turbine, expansion is done to get outlet stream at 95°C and 3.5 bar.

The stream after expansion in turbine is used as heat source in distillation column which is operated at 14 bar. Here distillation is the assembly of heater and separator to separate vapor and liquid. The main heat source is utilized in evaporator to provide desired stream at high temperature and pressure of 168°C and 20 bar respectively. The thermal efficiency reported are comparatively less of order of 20%, which is about 40% of the Carnot efficiency.

Some patent claim of using higher temperature heat source of order of 400°C. The vapors entering the turbine is at temperature of 350°C and 50 bar which is expanded to 1 bar pressure and 84°C. Mass fraction of vapors used is 0.5 with these condition efficiency achievable is reported as 54% of Carnot efficiency (Patent 4489563).

2.2.8.2.3   Ammonia Water Cycle as Bottoming Cycle with Steam Power Cycle

This patent reports of using two working fluids in same cycle as shown in Figure 2.6. Primary fluid used is steam which is expanded in turbine. The heat of stream obtained after passing through the turbine is utilized for heating multicomponent working fluid (i.e., ammonia-water mixture) in a separate closed loop.

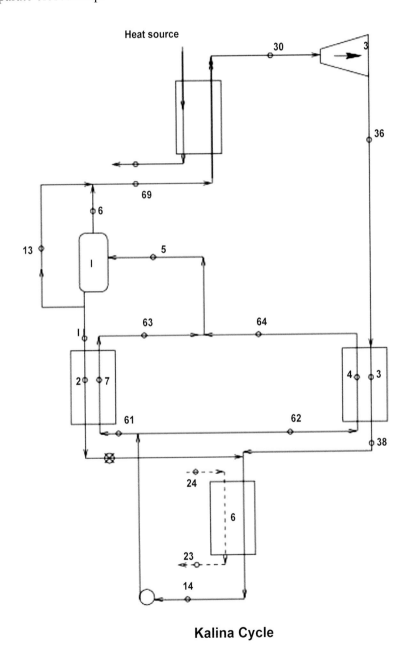

FIGURE 2.5 Low temperature heat source application.

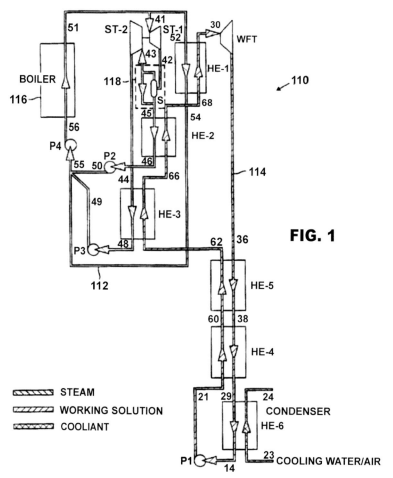

FIGURE 2.6 Ammonia water cycle as bottoming cycle to steam cycle.

The primary fluid in vapor state is expanded in first stage expander to obtain useful energy with partial condensation. Outlet stream from first stage expander is separated into vapor and liquid in a separator. This separated vapor stream is again expanded in second expander. After second expansion outlet stream exchanges heat with distill out ammonia at higher pressure. This vapor rich in ammonia is expanded in turbine to get useful output. Here cycle involving ammonia-water system is Bottoming cycle.

This expanded vapor of multicomponent fluid is condensed in condenser. The heat of expanded vapors of multicomponent fluid is recovered in recuperator by preheating the condensed liquid from condenser as shown in

Solar Energy for Green Engineering

Figure 2.6 This cycle uses heat source of 220°C. The steam to first turbine is at 210°C, 19 atm and is expanded to 140°C and 4 atm. In the second turbine, consider expansion from inlet condition of 140°C, 4 atm to outlet condition of 57°C, 0.17 atm. Bottoming cycle using ammonia-water as working fluid has expansion in turbine from 207°C, 40 bar to 128°C, 9 bar. This handles vapor mass fraction of 0.71. This patent claim thermal efficiency of order of 68% of Carnot efficiency (Patent 5822990).

2.2.8.2.4 Low Source Temperature and Low Pressure (LP) Use for Power Generation

This patent claims of using variable low temperature source from 110°C to 77°C (point 1 and point 6) as shown in Figure 2.7. Here the rich vapors are expanded in turbine from 2.5 bar, 103°C to 0.7 bar, 74°C. The composition of vapors used for expansion is 0.495 by mass fraction. After expansion, the heat of exhaust vapor stream is recovered with liquid stream going to vaporizer. The vapors are then condensed in condenser and pumped to a vapor liquid separator via a set of heat exchangers. The condenser is operated at 25°C. The vapors from vapor liquid separator are heated to higher temperature required at inlet of turbine and complete the cycle. The thermal efficiency reported is 50% of Carnot efficiency (Patent 5029444).

1. The advantages of Kalina cycle are as follows:

- The use of ammonia water mixture results in better thermal match in the boiler due to variable temperature boiling.
- Use of mixture enabled better internal heat recovery.
- Heat recovery at discharge of turbine has added heat recovery, and hence improved efficiencies are obtained.

EI-Sayed and Tribus (1985) showed that Kalina cycle has 10–20% higher second law efficiency than a steam Rankine cycle operating within same boundary conditions. A second law analysis of Kalina cycle has shown variable heating in boiler and reducing irreversibility because of efficient heat integration as cause of improved efficiencies (Park Sonntag Cycle Study, 1990).

1. Kalina cycle has a few disadvantages:
- Kalina cycle requires 100% vaporization of working fluid for expansion in turbine.

- The heat exchanger surfaces would dry out at higher vapor fraction, that would result in lower overall heat transfer coefficients and larger heat exchange area.

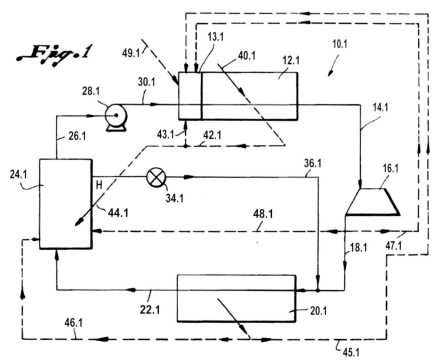

FIGURE 2.7 Low temperature and low pressure for power generation.

2.2.8.3 ROGDAKIS AND ANTANOPOLOUS CYCLE

They have replaced the distillation and condensation of Kalina cycle with absorption/condensation process as shown in Figure 2.8. Here rich NH_3/H_2O solution are heated in vapor generator, which produces vapor mixture of mass fraction of approximately 0.73 and weak solution of mass fraction 0.3.

The vapor is superheated and expanded in high pressure, MP, and LP turbine with intermittent reheating. The heat of liquid going out of vapor generator is recovered by heat exchange with the rich liquid fed to vapor generator. The high pressure side is maintained at 80 bar whereas the LP side is at 2 bar.

Solar Energy for Green Engineering

After LP turbine, heat is recovered in vapor cooler and condenser respectively and it is offered at high condensation temperature to preheater. In absorber, vapors with high mole fraction of ammonia are absorbed by weak liquor from boiler. The mixture is heated first with heat produced during absorption, then gradually heated by heat offered by condenser and vapor cooler, and is finally heated in vapor generator.

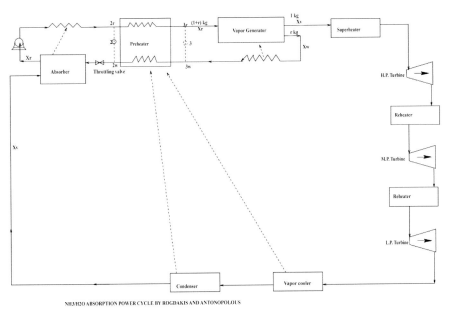

FIGURE 2.8 Antanopolous and Rogdakis cycle.

The heat source used here is at 400°C and heat sink is at 19.8°C. This cycle includes a very good heat recovery compared to other cycles reported in literature. The thermal efficiency achieved is reported as 65% of the Carnot efficiency achievable [19].

2.2.8.4 COMBINED POWER/COOLING CYCLE

This includes the idea of expansion to very low temperature without condensation. This very low temperature ammonia produces refrigeration. The low temperature ammonia is condensed by an absorption/condensation process. Net power output will be combination of power and refrigerant output. As shown in Figure 2.9, ammonia-water mixture (state 1) is pumped to high pressure (state 2). The mixture is heated to boil off ammonia (state 5) and vapor is enriched in ammonia by condensing part of vapors in condenser (state 6). The condensate richer in water is returned to boiler. Ammonia is superheated after the condenser to raise its temperature. The superheated vapor is almost pure ammonia as it is to be expanded in turbine to exit at very low temperature.

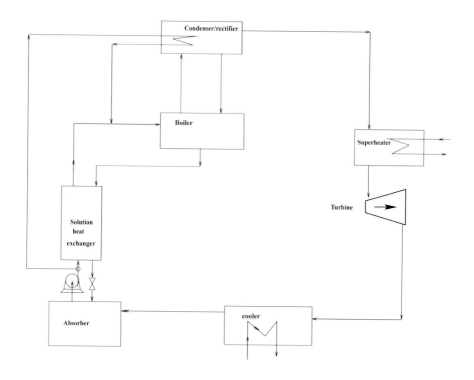

A MODIFIED AMMONIA BASED COMBINED POWER/ COOLING CYCLE

FIGURES 2.9 Combine power and cooling cycle.

After expansion through the turbine to generate power, low temperature vapors first provides cooling in the cooler and then absorbed by weak solution from boiler in an absorber, to form strong ammonia-water liquid solution and hence complete cycle. This cycle has reported to operate with the heat source of 137°C and heat sink of 7°C. The vapor composition at the turbine outlet is 0.99 (mole fraction). The thermal efficiency obtained

Solar Energy for Green Engineering

with the given configuration is reported 74% of the Carnot efficiency (Feng Xu et al., 2000).

2.3 CONCEPT OF MULTICOMPONENT FLUID BASED POWER GENERATION FOR EFFECTIVE CONVERSION OF SOLAR ENERGY TO POWER

2.3.1 MULTI-ABSORPTION LIQUOR AMMONIA ABSORPTION HEAT ENGINE (MALAE CYCLE)

The MALAE cycle proposes high quality of ammonia concentration at turbine inlet that allows expansion in turbine to low temperature. The cycle considers some percentage condensation (maximum allowable limit 10%) at turbine outlet that adds to the power output and hence improves the thermal efficiency. Figure 2.10 shows the proposed MALAE cycle. The rich ammonia-water mixture is pumped (state 4) to high pressure through two heat exchangers to distillation column as feed (stage 7). The rich liquid is vaporized in reboiler to give vapors (stage 14) that consists of majority of low volatile component, i.e., ammonia. Further enrichment of vapors occurs in distillation column without expense of additional heat. The partial condenser causes further purification of ammonia by partially condensing vapors (stage 12). There are a total of six stages for vapor-liquid contact in distillation column. Fifth stage receives the ammonia vapor and rest stages (except reboiler) act as the enrichment stages without any expense of energy. Parts of vapors that are condensed in partial condenser are sent back to column as reflux. Hence, vapors rich in ammonia (stage 1) are obtained at top and possible lean ammonia-water mixture (stage 15) is obtained at bottom. The vapors enriched in ammonia (stage 1) are superheated in superheater using the hot bottom of the distillation column, hence offers effective utilization of heat source. The vapors (stage 2) are expanded in turbine to low fixed pressure of 5 atm, with some percentage condensation of the ammonia-water mixture (stage 3). The enthalpy drop gives power output from turbine.

The vapors and liquid mixture past turbine and the lean liquid after heat exchange (stage 11) are used to regenerate the strong solution in the absorber with rejection of heat from cycle. The lean liquid from reboiler exchanges heat in superheater, solution heat exchanger 1 and solution heat exchanger 2, that bring best utilization of available heat. This split of heat exchangers has the added advantage for heat recovery. The rich liquid from absorber-condenser (stage 4) recovers heat from lean liquid from reboiler and additionally in the partial condenser.

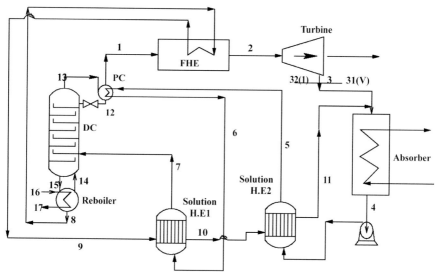

FIGURES 2.10 MALAE cycle schematic.

2.3.2 VAPOR-LIQUID EQUILIBRIUM

Calculation of vapor-liquid equilibrium for ammonia water becomes crucial during the designing of binary power plants. Calculation of respective composition of ammonia-water mixture for vapor and liquid mixture at respective temperature and pressure is crucial. Calculation of enthalpies of gaseous and liquid mixture depending on composition based on respective temperature and pressure need to accurate.

Modified Roult's law is used for calculation of liquid and vapor mole fraction, where the Virial equation of state and Redlich Kwong equation of state combinely is used for vapor non-idealities and Redlich Kister equation is used for calculation of non-idealities, i.e., activity coefficient for liquid phase. Basic fundamental equation with departure function from ideality is used for calculation of enthalpy and entropy. Gillespi experimental data is used to develop equation for calculations of vapor and liquid mole fraction (El-Sayed et al., 1985) [20, 21].

Solar Energy for Green Engineering

2.4 ANALYSIS OF MULTICOMPONENT/BINARY POWER CYCLE FOR DIFFERENT TEMPERATURE SOURCE USING SOLAR ENERGY

2.4.1 SIMULATION ANALYSIS

2.4.1.1 EFFECT OF NUMBER OF PLATES ON REBOILER LOAD IN DISTILLATION COLUMN

The use of multi-stage distillation column over the single-stage distillation is the unique contribution of the present cycle. Figure 2.11 shows a reduction in the heat load of reboiler with an increase in the number of stages in the column. It depicts reduction of approximately 13.5% of heat load compared to single stage heating. This advantage has enabled improved efficiency in the cycle.

Here heat is supplied only to first plate and further enrichment of vapor is obtained in the column with the help of additional stages without expense of any additional heat and hence reduces the overall heat load. The reflux requirement of column has also observed to be diminished causing reduction in reboiler load compared to less number of stages. No significant reduction in heat load was not observed after 6 number of stages, hence 6 number of stages were considered for further simulations.

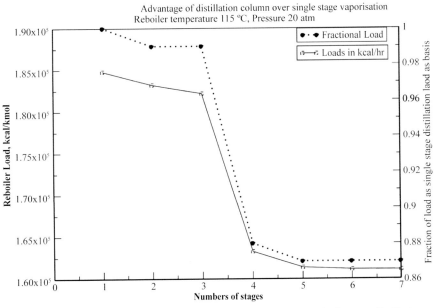

FIGURE 2.11 Advantage of multi-stage distillation column over single stage distillation.

2.4.1.2 EFFECT OF REBOILER TEMPERATURE

The optimal thermal efficiency has been observed to rise with increment in reboiler temperature. The increase in reboiler temperature results in decline of vapor flow-rate to turbine. Higher reboiler temperature give rise to higher superheater temperature, this causes expansion in turbine to lower temperature leading to more enthalpy drop, i.e., power output hence gain in thermal efficiency was observed.

2.4.1.3 EFFECT OF LOW PRESSURE (LP) ON THERMAL EFFICIENCY

The effect of LP in cycle on thermal efficiency is studied as shown in Figure 2.12. The reduction of LP in cycle, i.e., absorber pressure observed to improve the absorption giving rise to increment in absorber outlet composition. Richer absorber outlet composition, i.e., feed composition to column leads, reduction in reboiler load but also reduction of highest operational achievable pressure in distillation column. Low high pressure achievable in column results in reduction of turbine power output, hence thermal efficiency. Figure 2.12 shows optimized LP side to be 5 atm, as there is no significant increase in highest thermal efficiency achievable after this. The variation of pressure is studied from 1.8 atm to 5.5 atm. The increment in optimized thermal efficiency was observed from approximately 13 to 15.

2.4.1.4 EFFECT OF TURBINE INLET PRESSURE

The distillation column is operated at the turbine inlet pressure. Figure 2.12 shows the effect of pressure on the thermal efficiency of the power cycle. It shows that the efficiency of the power cycle goes through maximum with increase in the turbine inlet pressure. Increment in turbine inlet pressure causes increase in the pressure ratio across the turbine. There is increment in enthalpy drop across the turbine with pressure ratio, subsequently increasing the efficiency. However, the vapor production from distillation column goes down with increase in turbine inlet pressure as shown in Figure 2.13, which causes reduction in efficiency. Enthalpy gain from the increase in the pressure ratio makes up for the drop in the vapor flow-rate. However, at optimum pressure, the enthalpy gain from the higher pressure ratio could not make up for the drop in the vapor flow-rate and hence the efficiency declines.

Solar Energy for Green Engineering

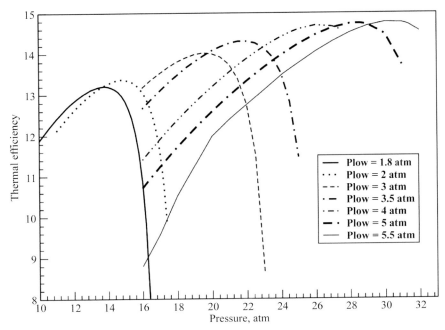

FIGURE 2.12 Effect of low pressure on efficiency.

2.5 CONCLUSION

The overview of different power cycles from traditional Rankine cycle, organic Rankine cycle using simple working fluid was presented. The binary fluid, e.g., ammonia water could be used to utilize the temperature effectively is presented. Kalina used ammonia water extensively at higher temperature heat source to produce power effectively. Ammonia water based cycle that can utilize the low temperature heat source effectively is presented. The novel power MALAE cycle was found to utilize low temperature heat sources using solar energy by producing steam @ 120°C, effectively using ammonia water as working fluid. The simulation has shown that cycle can achieve maximum efficiency of 14.74% at 115°C of reboiler temperature that operates with 120°C solar steams as heat source. Novel Cycle has demonstrated recovery of 72% compared to maximum that can be extracted given by Carnot cycle. Present novel cycle has demonstrated highest power extraction compared to maximum possible for low temperature heat sources concluding the very efficient heat integration.

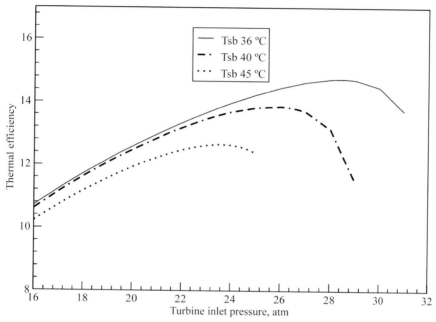

FIGURE 2.13 Effect of turbine inlet pressure.

KEYWORDS

- exergy
- green engineering
- Kalina cycle
- multicomponent mixture
- organic Rankine cycles
- power cycle
- Rankine cycle
- vapor-liquid equilibrium

REFERENCES

1. Kalina, A. I., (1984). Combined cycle system with novel bottoming cycle. *J. Eng. Gas Turb. Power, 106*(4), 737–742.

Solar Energy for Green Engineering

2. Goswami, D. Y., (1995). Solar thermal power: Status of technologies and opportunities for research. In: *Proc. of the 2nd ISHMT-ASME Heat and Mass Trans* (pp. 57–60). Conf., India New Delhi: Tata McGraw Hill.

3. Ibrahim, O. M., & Klein, S. A., (1996). Absorption power cycles. *Energy, 21*(1), 21–27.

4. Goswami, D. Y., (1998). Solar thermal power technology: Present status and ideas. *Energ. Source, 20*, 137–145.

5. Kalina, A., (1985). *Method of Generating Energy*. U.S. Patent: 4,548,041.

6. Ibrahim, O. M., & Klein, S. A., (1989). Optimum power of Carnot and Lorentz cycle. In: *Simulation of Thermal Energy System: Presented at the Winter* (pp. 91–96). Meeting of the American Society of Mechanical Engineers, San Francisco, California.

7. Kalina, A. I., (1984). *Generation of Energy*. U.S. Patent: 4,489,563.

8. Ei-Sayed, M., & Tribus, M., (1985a). *Thermodynamic Properties of Water-Ammonia Mixtures Theoretical Implementation for use in Power Cycle Analysis*. Center for advanced engineering study, Massachusetts Institute of Technology, Cambridge, MA, Draft report.

9. Bhagwat, S. S., Satpute, S. R., Patil, S. S., & Shankar, R., (2011). U.S. Patent: 8,910,477.

10. Ibrahim, A. H., (1997). Second law analysis of the reheat-regenerative Rankine cycle. *Energy Conversion and Management, 38*(7), 647–657.

11. Smith, Ness, H. C. V., & Abbott, M. M., (2004). *Introduction to Chemical Engineering Thermodynamics* (6thedn., Vol. 4). McGraw Hill.

12. Badr, O., Naik, P. S., Callaghan, W. O., & Probert, S. D., (1991). Expansion machine for a low power output steam Rankine-cycle engine. *Applied Energy, 39*(2), 93–116.

13. Boretz, J. E., (1986). Supercritical organic Rankine engines (SCORE). *Proceedings of the 21st Intersociety Energy Conversion Engineering Conference* (pp. 2050–2054). ACS Washington DC.

14. Angelina, G., & Paliano, P. C. D., (1998). Multicomponent working fluid for organic Rankine cycles. *Energy, 23*(6), 449–463.

15. Demuth, O. J., (1980). Analysis of binary thermodynamic cycles for a moderately low temperature geothermal resource. *Proceedings of the 15th Intersociety Energy Conversion Engineering Conference* (Vol. 1, pp. 798–803). AIAA, NY.

16. Demuth, O. J., (1981). Analysis of mixed hydrocarbon binary thermodynamic cycles for a moderate temperature geothermal resources. *Proceeding of the 16th Intersociety Energy Conversion Engineering Conference* (Vol. 2, pp. 1316–1321). IEEE, NY.

17. Ibrahim, O. M., & Klein, S. A., (1996). Absorption power cycles. *Energy, 21*(1), 21–27.

18. Martson, C. H., (1990). Parametric analysis of the Kalina cycle. *Journal of Engineering for Gas Turbine and Power, 112*, 107–115.

19. Rogdakis, D. E., & Antonopoulos, K. A., (1991). A high efficiency NH3/H20 absorption power cycle. *Heat Recovery System and CHP, 11*(4), 263–275.

20. Park, Y. M., & Sonntag, E. R., (1991). *Thermodynamic Properties of Ammonia Water Mixture: A Generalized Equation of State Approach* (p. 97). ASHRAE Transactions.

21. Ibrahim, O. M., & Klein, S. A., (1992). *Thermodynamic Properties of Ammonia Water Mixture*. Solar energy Laboratory, University of Wishconsin, Madison, WI.

22. Ibrahim, O. M., & Klein, S. A., (1996). Absorption power cycles. *Energy. 21*(1), 21–27.

23. Bhagwat, S. S., Satpute, S. R., Patil, S. S., & Shankar, R., (2013). *Improved Thermodynamics Cycle*. U.S. Patent: 8,910,477.

24. El-Sayed, Y. M., & Tribus, M., (1985). A theoretical comparison of the Rankine and Kalina cycles. Proceedings of Analysis of Energy Systems, Design and Engineering (pp. 185–211).
25. Park, Y. M., & Sonntag, R. E., (1990). A preliminary study of the Kalina power cycle in conjunction with a combined cycle system. *International Journal of Energy Research, 14*, 153–162.
26. Kalina, A. I., (1986). *Method and Apparatus for Implementing a Thermodynamic Cycle Using a Fluid of Changing Concentration.* U.S Patent: 4, 586, 340.
27. Kalina, A. I., (1988). *Method and Apparatus for Implementing a Thermodynamic Cycle with Recuperative Preheating.* U.S. Patent: 4,763,480.
28. Kalina, A. I., (1990). *Method and Apparatus for Thermodynamic Cycle.* U.S. Patent: 4,099,545.
29. Kalina, A. I., (1991). *Method and Apparatus for Converting Low Temperature Heat to Electric Power.* U.S. Patent: 5,029,444.
30. Kalina, A. I., & Rhodes, L. B., (1998). *Converting Heat Into Useful Energy Using Separate ClosedLoop.* U.S. Patent: 5,822,990.
31. Larson, D. L., (1987). Performance of the Coolidge solar thermal electric power plant. *Journal of Solar Energy Engineering, 109*(1), 2–8.
32. Feng, X. D., Goswami, Y., & Bhagwat, S. S., (2000). A combine power/cooling cycle. *Energy, 25*, 233–246.
33. Clifford K. Ho & Brian D. Iverson, (2014). Review of high-temperature central receiver designs for concentrating solar power, *Renewable and Sustainable Energy Reviews, 29:* 835–846.
34. Hank Price, EckhardLupfert, David Kearney, Eduardo Zarza, Gilbert Cohen, Randy Gee, & Rod Mahoney, (2002). *Advances in Parabolic Trough Solar Power Technology, J. Sol. Energy Eng., 124*(2), 109–125.
35. Vignarooban, K., Xinhai Xu, & Wang, K., et al., (2015). Vapor pressure and corrosivity of ternary metal-chloride molten-salt based heat transfer fluids for use in concentrating solar power systems, *Applied Energy, 159*(1), 206–213.
36. Dennis L. Larson, (1987). Operational evaluation of the grid-connected Coolidge solar thermal electric power plant, *Solar Energy, 38*(1), 11–24.
37. Ei Sayed M., & Tribus M. (1985a). Thermodynamic properties of water-ammonia mixtures theoretical implementation for use in power cycle analysis. Center for Advanced Engineering Study, Massachusetts Institute of Technology, Cambridge, MA, Draft report.
38. Park., Y. M., & Sonntag, R. E., (1990). A preliminary study of the Kalina Power Cycle in conjunction with a combined cycle system. *International Journal of Energy Research, 14,* 153–162.
39. Feng Xu, D, Goswami Yogi, & Bhagwat Sunil, (2000a). A combine power/cooling cycle. *Energy, 25*, 233–246.
40. El-Sayed, Y. M., & Tribus, M., (1985). A theoretical comparison of the Rankine and Kalina cycles, *Proceedings of Analysis of Energy Systems, Design and Engineering,* 185–211.

CHAPTER 3

Characterization Techniques for Bio-Nanocomposites

SHRIKAANT KULKARNI

Assistant Professor, Vishwakarma Institute of Technology, Department of Chemical Engineering, 666, Upper Indira Nagar, Bibwewadi, Pune – 411037, India, Mobile: +91-9970663353, E-mail: shrikaant.kulkarni@vit.edu

ABSTRACT

The characterization at different scales or thorough investigations of nano-composites, consisting of one or more inorganic particles and biomolecules, demands a host of techniques. It involves the exploration of bio-nanocomposite materials at the atomic level, elemental composition, degree, and the kind of the chemical bonds that exist. Moreover, it refers to the determination of size and structure, other than the identification of interactions between the inorganic and biological components of the system. Further, specific properties of the materials are examined not only because a material exhibits a specific function but also it gives an insight to probe the inorganic-biological interface. Traditional techniques derived from the knowledge of chemistry, and material science and biology have been employed studying these complex bio-nananocomposite materials. Thus an overview of key analytical methods is presented, ranging from physical principles to practical aspects is taken which throws light upon various aspects of bio-nanocomposite materials.

3.1 INTRODUCTION

Thorough investigations of composites, consisting of one or more inor-ganic particles and biomolecules, demand the combination of a plethora of techniques, providing for its characterization at different scales. Presents information about the nature of chemical elements, and their concentration level, and the nature chemical bonding between them. Further, it throws light on the size, shape geometry, morphology, and structure of the bionano-composite, apart from the identification of nature of bonding between the inorganics and biomolecules. Finally, selective properties of the materials must also be studied to give an insight into specific function and to probe the inorganic-biological interface.

So far, no distinct or unique technique has been developed for the study of specific properties of bionanocomposites. Traditional methods based on wet chemistry, material science, and biology has been used to investigate these complex bionanocomposites. Key analytical methods, based on physical principles to practical aspects, are employed to explain their use in bionanocomposites.

3.2 CHEMICAL ANALYSIS OF POLYMERS

The purpose of chemical analytical techniques is to estimate the nature of the elements present in a given sample, their relative amount, and the chemical interactions between them. In principle, the composition of the reagents used for the synthesis and their chemical routes present an indication of various elements found in the ultimate product. However, unreacted species, impurities, or by products of the reaction may also be present that need to be identified. Thus, techniques that provides a simultaneous indication of all present elements and is better over other techniques which are confined to a single element, unless a precise titration of a given element is necessary.

3.3 MECHANICAL COMMINUTION, SEPARATION, AND IDENTIFICATION OF POLYMERS

Mechanical comminution: It involves use of with cutting edge like shears, knives, or razor blades and Drilling, milling, etc., such that a small particle diameter is derived. The repeatability and coherent nature of the comminution techniques is of paramount importance when subjected to mechanical loads or low temperature treatments.

Comminution, is followed by conditioning of samples Comminution, is followed by conditioning of samples over P_2O_5 at ambient temperature in a desiccators, e.g., plasticizers (1–2 g) can be extracted with diethyl ether by using Soxhlet apparatus. Stabilizers depending upon pure organic or organo-metallic substances can be separated partially. After distilling off the ether followed by subjecting the extract to drying at 105°C to fixed mass, the amount of components dissolved in ether is measured from the difference in mass before and after the extraction of the extraction flask. Plasticizers namely esters of some aliphatic and aromatic mono and dicarboxylic acids, phosphorus acid esters, ethers, alcohols, ketones, amines, amides, as well as non-polar and chlorinated hydrocarbons (HCs) [9].

Thin-layer chromatography (TLC) is preferred for their extraction and qualitative analysis. Usually Kieselgur plates, (0.25 mm thick, at 110°C for 30 min), with the vapor saturation is used. Methylene chloride and mixtures with diisopropyl ether at temperatures ranging from 40 to 60°C are so far successfully used as the solvents. TLC was preferred separation method characterized by its high resolution efficiency, rapidity, and variety of detection possibilities. Usually 0.5 mm thick silica-gel-G-plates are used, activated at 120°C for 30 min. in a supersaturated atmosphere. However, HPLC independently or HPLC-MS coupled techniques are used preferably over TLC because of several advantages of it [10].

Although selectivity of TLC can be increased with the optimization of process variables, gas chromatography should be used in preference for complex plasticizer formulations. The gas chromatography based separation of plasticizers can be performed with or without derivatization to substances with low boiling points by transesterification. Fillers can be determined quantitatively in polymers by centrifugation followed by decantation of the solvent. Carbon black as an exception cannot be separated totally, by using centrifugation and other specialized methods. Common methods of qualitative inorganic analysis can also be employed for filler identification. Quantitative determination is done gravimetrically [11].

3.4 QUALITATIVE AND QUANTITATIVE ANALYSIS OF POLYMERS

Polymer identification starts with a sequence of preliminary tests. Polymers are difficult to identify because of factors like presence of copolymers, macromolecular properties, etc., unlike low molecular weight compounds, which can be identified using physical constants like melting or boiling points, etc. Further physical methods like IR, NMR spectroscopy, as well as gas chromatography on pyrolysis will have an added advantage.

Pyrolysis involves about heating 0.1 g of the sample with care in a 60 mm in length glow tube, diameter of 6 mm over a tinyl flame. De-polymerization is a typical example of thermal breakdown which is observed with clarity namely in PVC and PVAC. De-polymerization, elimination, and statistical chain-scission reactions can be employed for polymer characterization. If the monomer is the major product of degradation, it can easily be identified from physical properties like boiling point and refractive index. Elimination and chain-scission reactions present distinct thermograms that can be studied using either gas chromatography or IR spectroscopy. For studying depolymerization tendency of a typical polymer sample (0.2 to 0.3 g) is heated gently to a temperature of 500°C in a distillation flask. The boiling point and refractive index of the fraction distilled, is then determined [12, 13].

3.5 QUALITATIVE AND QUANTITATIVE ANALYSIS OF STABILIZERS

Stabilizers like heat, light, and oxidation resistant have acquired a great importance. Different examples of stabilizers are shown in Table 3.1.

TABLE 3.1 Representative Examples of Stabilizers

Stabilizers	Heat Absorbers	UV Light Absorbers	Antioxidants
Examples	Heavy metal salts, and organic acid metal salts, nitrogenous organic compounds	Hydroxybenzophenone–derivatives, salicyl esters, and benzotriazoles	Phenols, aromatic amines, and benzimidazoles

Qualitative analysis of stabilizers is cumbersome because of their huge number and amounts in traces, vulnerability to undergo transfer or rearrangement reactions their detection is of importance because of hazardous nature of its decomposition products. Stabilizers are identified either by solid liquid extraction or by precipitation from a dilute solution [14].

3.6 SPECTROSCOPIC TECHNIQUES

3.6.1 INDUCTIVELY COUPLED PLASMA (ICP)

The analysis of solid objects can be carried out by bringing about degradation of the sample and recovering the components of the solid in solution (by dissolving it in acidic medium) or in the gas phase (by using laser ablation).

Inductively coupled plasma (ICP), technique involves the use of an inert gas matrix at higher temperature (up to 80,000 K) which allows the ionization of the

Characterization Techniques for Bio-Nanocomposites 53

atoms or molecules present in the sample in either dissolved or vaporized state. The ionized species can then be detected by mass spectroscopy (ICP-MS) [1], or the light emitted because of the recombination of electrons and ions which can then further be analyzed by atomic emission spectroscopy (ICPAES) [2].

These tools are very efficient in the estimation of the elements quantitatively and simultaneously, provided interferences between them are regulated. However, since the sample is degraded, information pertaining to their chemical interactions is difficult to ascertain.

3.7 INDUCTIVELY COUPLED PLASMA (ICP) ATOMIC EMISSION SPECTROSCOPY

Principle: When the sample is vaporized, atomized by collision at an elevated temperature, the generated plasma energetically promotes argon species to emit radiation of characteristic wavelengths.

3.7.1 THEORY

ICP atomic emission spectroscopy (ICP) technique involves, vaporization of the sample the element of interest is atomized at an extremely elevated temperature (7000°C), argon plasma, produced, and maintained using radio frequency waves. The atoms on collision with energetically excited argon species emit characteristic spectra which are detected by a detector like photomultiplier tube (PMT).

The separation of spectral lines is brought about in two ways:

- A sequential or scanning ICP: a scanning monochromator having a moving grating is made use of to bring the monochromatic light of the target wavelength to a detector; and
- A simultaneous or direct reader ICP: a polychromator possessing a diffraction grating is employed to disperse the light into varied wavelengths. Either water (with 10% acid), or an appropriate organic solvent like xylene [39].

3.7.2 ADVANTAGES OF ICP

- Better detection limits, and a wide linear range for many elements.
- A direct reading instrument is useful in multi-element analysis at extremely fast pace.

- No reference is required as all samples are converted into water or organic solvent soluble matrices prior to analysis.
- Interferences (chemical and ionization) are contained in ICP analysis as against AAS wherein they are frequently observed.
- A scanning instrument allows to go to an interference free position.

3.7.3   LIMITATIONS OF ICP

- ICP instruments are constrained by the analysis of samples in liquid state only.
- Solid samples are to be dissolved in a suitable solvent prior to analysis.
- The volume of sample solution required for analysis has to be minimum 25 ml.
- demands extensive sample preparation methods.
- ICP instruments are not so rugged.
- Continuous monitoring while introducing the sample and torch systems, is necessary.
- Spectral interferences due to presence of some major metals may hinder the estimation of elements in traces which ask for the corrections for accurate quantitative determination [40].

3.8   FOURIER TRANSFORM INFRARED SPECTROSCOPY

Principle: The principle followed here is the structural information can be ascertained based upon the characteristic vibrational frequencies absorbed by the molecule which depends upon vibrational transitions it undergoes.

3.8.1   THEORY

Fourier transform infrared spectrometry (FTIR) is a special technique. It is employed to characterize samples which are present either in fewer amounts or as a small thing. Gels embodied in a rubber sample, need to be microtomized (i.e., chopped off into very thin slices) and deposited onKBr plates. Samples laced with inorganic compounds are usually characterized by both X-ray and FTIR Microscopy. Sample size is 20 µ can be tested by the FTIR-microscopy. FTIR is normally employed for qualitative estimation of several functional groups. Quantitative analysis by FTIR demands the application of suitably characterized references. In few examples, the peak absorbance is comparatively low, therefore a thicker film, 6 mm, is used.

Vibrational transitions (mid-IR) are quite useful for most of the studies. They offer information pertaining to either the presence or absence of typical functional groups in a given sample matrix. In practice, all functional groups which are infrared active show that fundamental band in a very narrow region of wavelength in the mid-IR region. Further, the complete spectrum, having fundamentals, overtones, and combination bands, shows a unique fingerprint region for the sample in question. It means that although we don't know anything about the sample, we will come to know about it in the later stage if it recurs. Finally, the absorption intensity of any given band, fundamental or overtone is dependent upon the number of functional groups generating the signal [49].

Modern dispersive spectrometers are double beam ones which chops off the incident beam into two parts; one beam passes through the sample and the other passes through an appropriate reference. The intensity of both the beams is observed with a detector, and the output is displayed as transmittance or absorbance:

$$\text{Transmittance} = 1t/I_0$$
$$\text{Absorbance} = -\log 10/I_t$$

where, I_t, I_0 refer to the intensity in the sample of reference, transmitted, and incident beam, respectively. The samples characterized are as follows:

A. **Gases:** 250 cm^3 at 1 atm;
B. **Liquids:** 0.25 cm^3;
C. **Solids:** 1 mg to get a spectrum;

The trace analysis in samples are required and more samplesare required proportionally.

The advantages of this technique are:

- Fast and cheap;
- Quite specific to some functional groups;
- Quite sensitive to some functional groups;
- Unique fingerprint analysis.

The disadvantages of this technique are:

- Demands typical cells, made up of NaCl, KBr, quartz, etc.;
- Normally requires dissolved samples;

56   Green Chemistry and Green Engineering

- Difficulty in getting better quantification in solids;
- Should calibrate all signals.

Interferences may occur from:

- Water interferes almost all IR work in reality;
- Solvents normally interfere and should be used selectively with care;
- In multi-component sample analysis species mutually interfere one another. Separations are required frequently;
- Optical components interfere too to various degrees in various regions, e.g., quartz is good for UV/Vis/Near-IR, but poor for mid-IR/far-IR [50].

3.9   ATTENUATED TOTAL REFLECTANCE (ATR) SPECTROSCOPY

Principle: The basic principles are the same as UV-visible spectroscopy that is, irradiating the sample with a light at different wavelengths and recording the outcoming light over the same spectral domain.

3.9.1   THEORY

Infrared (IR) spectroscopy technique is employed commonly for the identification of the chemical interactions in a given molecule base on the vibrational frequencies of the connected atoms. The decrease in light intensity at specific wavelengths corresponds to a mode of vibration in the given molecule which absorbs. The range of wavelengths employed in IR spectroscopy is 2 and 20 m unlike UV-visible spectroscopy for which it is 200, 800 nm. Apart from absorption, reflection, and scattering effects of light are likely to be witnessed in solids, thereby stifling the complete transmission of the radiation through the sample on absorption.

In IR spectroscopy, the attenuated total reflectance (ATR) technique is most commonly used [3]. The principle underlying this technique is generation of evanescent waves on the surface of a dense solid in air when it is subjected to irradiation at an angle leading to the total reflection of the incident beam. Therefore, the intensity of the light transmitted by the solid is attenuated as against the intensity of incoming light. A solid or a liquid coating on the given surface alters the spectral characteristics of the generated evanescent waves and consequently that of the outcoming beam that can be examined.

Characterization Techniques for Bio-Nanocomposites 57

3.9.2 APPLICATIONS

The ATR technique doesn't require sample preparation specifically. However, precaution has to be taken so as to maintain a better contact of the surface of the sample with the incident beam of light from the source. The extent, to which the evanescent wave generates above the dense solid surface, is a measure of the depth to which light penetrates within the given sample, which is of the order of 1 m. The analysis, therefore, is brought about at bulk level.

3.10 X-RAY PHOTOELECTRON SPECTROSCOPY (XPS)

Principle: XPS working depends upon the photoelectric effect giving rise to the ejection of electrons from deep energy levels of atoms (core and valence levels sometimes) on irradiating it with X-rays which are collected and analyzed.

3.10.1 THEORY

Bionanocomposites are characterization area corresponds to the very surface of the material, that is, at the inorganic biological interface. The contribution made by the chemical elements and groups in this interface to the whole signal is enhanced as against the bulk contribution. This makes x-ray photo-electron spectroscopy (XPS, also formerly named ESCA) is very useful technique for studying surface topography of solids [4].

XPS works by using the photoelectric effect principle explaining the removal of electrons from the energy levels deep enough of atoms (core and sometimes valence levels) on irradiation with X-rays. The photoelec-trons emitted from the surface of the sample with a given kinetic energy are trapped and analyzed. During their migration to the surface, the electrons can interact with other atoms, thereby reducing their kinetic energy. Therefore, it is assumed that only the first 5 nm thickness of the sample surface is analyzed. Since buried atomic energy levels are explored, many of the elements will have a specific and narrow kinetic energy range. Many peaks can be seen, showcasing variations in the oxidation, protonation state, and/or chemical environment of the element in question. Further, the relative peak area is the measure of the elemental concentration of the examined surface of the solid, when other factors such as the nature of the sample and characteristics of the equipment too are given due consideration. This demands to develop a

58 Green Chemistry and Green Engineering

calibration curve using reference materials before beginning with a sequence of experiments.

3.11 AUGUR ELECTRON SPECTROSCOPY (AES)

Principle: The electrons generated due to auger effect, i.e., post photoionization process are analyzed for the characterization of sample is the principle of auger electron spectroscopy (AES).

3.11.1 THEORY

It is very vital to mention that the photoionization process leads to many other complex phenomena, resulting into comparatively weaker but informative peaks. The process which brings about the return of ion to its neutral state leads to the ejection of an electron from some other level. This process is referred to as the Auger effect. It is a technique that is not so common unlike XPS but can complement largely to it [4].

The XPS requires samples in a dry state as the measurements are undertaken under high vacuum of the order of 8 to 10 mbar. The sample is normally mounted in place on a holder with a double sided adhesive tape. Similar to other surface analysis techniques, either the film deposited or the powder has to be as flat as possible. However, agglomeration of nano powders, can affect their XPS response [5]. As XPS probes the very surface of the material, it is highly sensitive to contamination in particular during preparation of sample. In majority of the cases, carbon impurities in traces get detected. Considering bionanocomposites the recent development of near ambient pressure (NAP)-XPS technique is most important, providing for recording of the spectra at a pressure of 20 mbar. This is a breakthrough development as it offers the exploration of solid sample surfaces in the presence of liquids, e.g., it is possible to study the effect of pH on amino acids deposited on the surface [6].

3.11.2 APPLICATIONS

NAP-XPS demands a high energy beam obtained from synchrotron radiation, and is applicable to:

- Characterization of more complex materials like for the analysis of inorganic-organic interfaces in bionanocomposites [7].

Characterization Techniques for Bio-Nanocomposites 59

- The whole XPS spectrum recorded for all samples, show peaks over a broad energy range which corresponds to C, O, Si, S, N, and Cl. It is possible to determine quantitatively the amount of each element.

3.12 X-RAY FLUORESCENCE SPECTROMETRY (XRF)

Principle: The binding energy of photoelectrons ejected by the sample on irradiating it with X-rays is measured for ascertaining the information about chemical composition, etc. XRF is based on the principle that each element emits its own characteristic X-ray line spectrum.

3.12.1 THEORY

XPS is a modern tool for the characterization of carbonaceous materials in terms of the chemical composition, impurity presence, and nature of chemical bonds. The XPS analysis performed on the CNT surface is used to measure the binding energy of photoelectrons ejected when CNTs are irradiated with X-rays [57, 58].

X-ray fluorescence spectrometry (XRF) is a non-destructive method of elemental analysis. When an X-ray beam is incident upon a target element, orbital electrons are dislodged. The vacancies or holes generated in the inner shells are filled by outer shell electrons. Energy releases during the process in the form of secondary X-rays known as fluorescence (FLU). The energy of the emitted X-ray radiation is fallout of the distribution of electrons in the excited atom. A unique electron distribution of every element help produce the quantitative analysis of Ba, Ca, Zn, P, and S in additives and lubricating oils, lead, and sulfur in gasoline, sulfur in crudes and fuel oils, and halogens in polymers. The high analytical precision of wavelength dispersive X-ray fluorescence spectrometry (WDXRF) has made it possible to develop methods for the precious metal assay of catalyst used in reforming process as against the precision of classical wet chemical methods. Metals like Pt, Ir, Re, or Ru can be been determined from numerous catalysts.

3.12.2 APPLICATIONS

The most important application of XRF in nanocomposites (NCs) is:

- It is nondestructive and requires minimal sample preparation.
- To identify the chemical composition of the modified carbon nanotubes surface.

60 Green Chemistry and Green Engineering

- To identify the presence of impurity.
- Type of chemical bonding.
- The elemental composition (qualitative) of unknown materials like engine deposits can be determined and provides the information in developing the sample preparation methods before analysis by inductively coupled plasma atomic emission spectroscopy (ICPAES) is undertaken.
- WDXRP, when properly calibrated, offers precision and accuracy comparable to wet chemical methods of analysis.
- EDXRP offers rapid qualitative analysis of unknown materials.
- Computer programs are available to correct for all of kinds of interferences.

Alternatively, a semi-quantitative analysis can be provided by XRF alone, in nondestructive testing when a small amount of sample is available. The deposit need not be removed from the piston since large objects can be placed directly inside an EDXRF spectrometer. Aqueous and organic liquids, powders, polymers, papers, and fabricated solids can all be analyzed directly by XRF. Although the method can be used to analyze materials varying in size from milligram quantities to bulk parts like engine pistons, minimum 5 grams sample is normally required for accurate quantitative analysis.

Due to many inter elemental matrix effects, matrix matched standards apart from a blank are required for quantitative analysis accurately. The detection limits for XRF are comparatively higher than other spectrometric methods and reasonable drop in sensitivity is observed in particular for light elements such as magnesium. In the analysis of solids, particle size and geological effects can play an important role.

The most common interferences are:

- Absorption and/or enhancement of the element of interest by other elements in the matrix;
- Line overlaps.

3.13 ENERGY DISPERSIVE X-RAY SPECTROSCOPY (EDS)

Principle: The radiation interacts with the core levels of the atoms, the energy of the emitted radiation is the characteristic of an element, and thereby allows for its characterization.

3.13.1 THEORY

The major limitation of XPS is that it can't be straightway coupled to an imaging technique when surface analysis is to be done. The energy dispersive x-ray spectroscopy (EDX or EDS or EDXA) is most commonly and conveniently used in such cases. When coupled to electron microscopy (scanning electron microscopy (SEM) or transmission electron microscopy (TEM)), it works on the principle of the ability of the electron beam to promote electrons from the core levels to a higher energy level of a given atom and subsequent de-excitation, i.e., demotion of an electron from a high energy level to the partially depopulated core (lower) level, with the emission X-ray. Several photons can be detected, because many levels of the same atom can be involved in their generation. This process resemble with the Auger effect except that it leads to light emission and not electron removal.

3.13.2 APPLICATIONS

EDX is nowadays a routine technique for the qualitative chemical analysis of surfaces [8]. The advantage is as several peaks allow for elemental characterization, and increases the probability for overlaps in energy for different elements, which can be major cause of concern during analysis. Use EDX for quantitative analysis is however difficult. Light elements (such as C, N, and O) don't have emission characteristics favorable enough and therefore their accurate estimation is debatable.

Further, it is difficult to perform the metallization step usually carried out for the characterization of nonconductive samples using SEM, as materials (fragile) may give way during the course of recording of the EDX spectra.

3.14 ELECTRON ENERGY LOSS SPECTROSCOPY (EELS)

Principle: Analysis of the influence of the material on the kinetic energy of incoming electrons rather than monitoring the effect of electrons on the material under examination.

3.14.1 THEORY

Electron energy loss spectroscopy (EELS) is gaining popularity over time as a substitute to EDX [9]. In this technique, the electron beam is used, so as to make it adaptable to coupling of EELS with microscopy (SEM and TEM).

There is a decrease in energy of the incoming electron beam on account of interaction with phonons, inter-band electronic transitions, or core level ionizations. These phenomena occur at energies that are characteristic of a given element, enabling its detection. Similarly a few other interactions can provide the additional information about the oxidation state and coordination tendency of the given atom or ion. In that respect, EELS compliments EDX and is adaptable to organics, including biological.

3.14.2 APPLICATIONS

EELS offer exciting applications in the domain of bio-functionalized nanomaterials (NMs) [10]. Magnetic iron oxide nanoparticles coated with MgO layer and further by a silica shell have been used for the purpose. E.g., Grafting of aminated silanes adsorb G-protein, which is followed by the introduction of the respective antibody. These particles were first examined using scanning transmission electron microscopy (STEM), which distinguishes core and shell. Sample cartography using EELS provides for the identification of the presence of Fe in the particle core, while O, and is found to be present both in the particles and the proteins. However, the presence of carbon and nitrogen at the surface of the particles, strongly justifying that adsorption has taken place.

3.15 CIRCULAR DICHROISM (CD) SPECTROSCOPY

Principle: Circular dichroism (CD) results because of the interaction between the polarized light and optically active, chiral, asymmetric molecules. CD is obtained by subtracting the absorption of left and right handed circularly polarized light (L and RCPL) over a broad range of wavelengths.

3.15.1 THEORY

The unraveling 3-D structures of proteins have become possible by X-ray diffraction, NMR spectroscopy, and cryo-electron microscopy. Although X-ray crystallography offers the detailed structural information, still it demands crystallization of protein under study. Cryo-electron microscopy on the other hand is especially useful for large protein complexes, which are aren't not easy to crystallize. Small proteins on the other hand are amenable to NMR analysis.

Although these techniques are very powerful tools for eliciting the detailed 3-D structure of biomacromolecules, they at times are heavy to bring about and are not amenable to study changes in conformation in response to environmental changes. Spectroscopic methods are a better option in that case. Though absorption and FLU spectroscopy very useful in studying the molecular changes, interpretation of such measurements is difficult in respect of structural changes (secondary). Therefore, techniques using polarized light are of vital importance.

A right handed circularly polarized light interacts in a different manner with a right-handed-helix than that with left-handed circularly polarized light thereby, CD signal may be either positive or negative, based on whether LCPL is absorbed to a more degree than RCPL (CD signal positive) or to a lesser degree (CD signal negative). Since most of the bio-molecules and their higher order structures are chiral, CD spectroscopy is a very powerful analytical tool in biology to characterize the structure and conformation of bio-molecules and to derive structural, kinetic, and thermodynamic data. CD is not an intrinsic property of the biological molecule simply.

Thus, the CD spectrum of a protein or of nucleic acids is not simply a summation of the individual residues or bases spectra but is largely affected by the 3-D structure of the macromolecule in question. Every structure has a unique CD signature, which is used to qualitatively analyze structural elements and to follow changes that take place in the structure of chiral macromolecules based on the environmental factors such as pH, temperature, concentration, solvent, cofactors, or denaturing agents, e.g., the most widely studied CD signatures for the characteristic secondary structural elements of proteins like the helix and the sheet which help in predicting the % contribution of each secondary structural element in the structure of a protein at large.

3.15.2 APPLICATIONS

CD spectroscopy tool is quite useful in exploring the effect of the interaction of biomolecules with inorganic nanoparticles, e.g., the influence of silica nanoparticles on the conformation of the T4 lysozyme molecule has been explored. [39]. As one of the leucine groups of this protein was expected to play a key role in its structural stability, mutants of leucine amino acid such as tryptophan (resulting into a decrease in stability) and cysteine (leading to an increase in stability) due to formation of disulfide bonds were used as substitutes. CD spectra of all three proteins are similar. However, silica nanoparticles, (9 nm diameter), structural changes were seen for all proteins, but the dominant form was the tryptophan mutant, in tandem with its low stability.

3.16 ISOTHERMAL TITRATION CALORIMETRY (ITC)

Principle: Isothermal titration calorimetry (ITC) is a very powerful tool for determining the thermodynamic, association constants, and stoichiometry of molecular interactions in aqueous solutions [40, 41] based on heat or change measurement during a physical or chemical change.

3.16.1 THEORY

Any chemical or physical change is associated with a change in heat or enthalpy; the calorimeter can be used as a universal detector. It helps in studying specifically the interactions of biomacromolecules with other molecules, or metal ions and the assembly of macromolecular complexes, and finally with nano-objects. ITC quantifies the heat liberated or the change in enthalpy (H) during a reaction. The stoichiometry and association constant (Ka) can be determined by titration mode. The system works by comparing thermocoupled sample and reference cells. Upon titration of a chelating agent or ligand in the sample cell, heat is either liberated (exothermic) or absorbed (endothermic), and the heat is measured based on the quantum of power required to maintain the temperature balance between sample and reference cells.

When the reactant saturates in the cell, the heat signal declines, and integration of the heats of each titration, helps in developing a binding curve. From the association constant the change in Gibbs free energy (G) can be calculated:

$$G = -RT \ln Ka$$

And with G and H the change in entropy (ΔS) can be calculated:

$$\Delta G = \Delta H - T \Delta S$$

3.16.2 APPLICATIONS

For bionanocomposites, ITC has been used to monitor the bonding between inorganic and biomolecules, e.g., to study the thermodynamic changes that accompany interaction of zinc oxide (ZnO) particles with ZnO binding

Characterization Techniques for Bio-Nanocomposites

peptides [42], collagen triple helices and the surface of silica particles [43]. The later has helped in identifying the specificity of collagen interactions, based on the surface chemistry of silica particles, and to reach particle surface saturation. Moreover, it's coupling with DLS, zeta potential measurements, and TEM very well showcases the contribution made by these characterization techniques in illustrating the engineering of bionanocomposites. The biomolecule inorganic interface at the molecular level presents a key for acquiring and commanding control over bionanocomposites design and development.

Other useful techniques like surface plasmon resonance (SPR) is based on the alteration in the plasmon band of a nanostructured gold thin film when a molecule is mounted on its surface and its counterpart is added to solution, and then their binding is optically detected. The major advantage of SPR is that it offers information about the kinetics of the binding process and permits to undertake desorption study. SPR supports largely to ITC which presents thermodynamic parameters, and stoichiometric quantitative data of the reaction, e.g., the interactions of particles of varying sizes and hydrophobicity with plasma proteins was studied by Cedervall et al. [44].

3.16.3   DETERMINING SIZE AND STRUCTURE IMAGING

The major limitation of the imaging technique is the resolution. It is defined as the smallest distance between two adjacent points, which are considered as separated. The resolution efficiency of a light microscope is contained by diffraction and is given by the formula:

$$r = \frac{0.61\lambda}{\mu \sin\alpha}$$

where, λ – wavelength of radiation; μ – refractive index; and α – semi-angle subtended at the specimen.

Electron microscopy technique has much better resolution which is attributed to the lower wavelength of electrons than photons. When electrons interact with a sample, resulting into a number of signals may be generated.

The traditional electron microscopes have many limitations which have their reflections in imaging of the specimens based on their nature. For example, the specimens necessarily have to be vacuum stable at ambient temperature and pressure, the path length of electrons is only a few nm), as well as the samples should be stable to the interaction of electron beam (i.e., neither thermally sensitive nor photosensitive).

3.17 MICROSCOPIC TECHNIQUES ELECTRON MICROSCOPY

Principle: Magnification and resolution of the samples under investigation using microscopic techniques for getting an idea about the particle size, morphology, etc.

3.17.1 THEORY

Thin sections (100–200 nm) and refractive indices differing about by 0.005 are mounted on glass slides are examined per se or with oil to rid-off microtoming artifacts, e.g., different layers in coextruded films, dispersion of fillers, and polymer grain size are determined [28].

Polymer blends (domain size: 1 pm) can be examined by using following optical microscopes based techniques:

- Polarized light when either the polymer phase is crystalline or inorganic filter and are agglomerated, e.g., nylon/EP blends with fillers like talc.
- Incident light for bulk sample surfaces, e.g., carbon black dispersed in rubber.
- Bright field for thin sections of samples loaded with carbon black, e.g., carbon black dispersed in thin films (TFs) of rubber materials.

However, scanning electron microscopy (SEM) and/or TEM are used when the domain size lies between 10 nm to 1 m [29, 30].

Samples by SEM can be tested for either:

- General morphology;
- Freeze fractured surfaces;
- Microtome blocks of bulk solid samples.

Contrast is obtained by a single or combination of the following methods:

- **Solvent Etching:** When there exists a large solubility difference in a given solvent of the polymers, e.g., PP/EP blends.
- **O5O4 Staining:** When at least 5% unsaturation in the polymers exists, e.g., NR/EPDM, BIIR/Neoprene.

Characterization Techniques for Bio-Nanocomposites 67

- **RuO4 Staining:** When neither solubility nor unsaturation differences exist, e.g., dynamic vulcanized alloys [31, 32].

3.18 SCANNING ELECTRON MICROSCOPY (SEM)

SEM can be used to:

- Examine liquids or polymers sensitive to temperature on a cryostage.
- Do X-ray/elemental analysis.

SEM technique is used for qualitative analysis. XRD analysis and mapping of the elements present is very useful for the analysis of inorganic fillers and their dispersion, inorganic contaminants in gels or on the surfaces and curatives qualitatively, e.g., aluminum, silicon, or sulfur in rubber and Cl and Br in halobutyls [33].

3.19 TRANSMISSION ELECTRON MICROSCOPY (TEM)

TEM is used when:

- A more in deeper study with domain sizes < 1 or so is carried out on polymer phase morphologies, e.g., dynamically vulcanized alloys, carbon black filler location in rubber products and morphology of block copolymers.
- Thin sections based on the nature of sample.
- Microtoming of the sample is done using a diamond knife near liquid nitrogen temperature ($-150°C$).
- The contrasting media is same as that for SEM.

3.19.1 SAMPLES PREPARATION FOR SEM ANALYSIS

Sample preparations methods for SEM include:

- freeze-drying; and
- critical point drying.

Both these techniques reduce structural damage due to surface tension effects because of air drying wet samples, e.g., the solvent exchange steps for the dehydration of samples before critical point drying may result into loss of structure or dissolution of certain components to certain extent. Although critical drying preserves the surface chemistry of a sample specimen, surface tension effects can damage it when liquid turns into a gaseous phase. However, at the critical point, materials can be transformed from liquid to gas, thereby avoiding structure damages without crossing the interfaces., e.g., with a critical point of water at 374°C, and 229 bar, biological samples can be dehydrated only by destroying samples. Therefore, liquid carbon dioxide (CO_2), with a critical point at 31°C and 74 bar is a better substitute than water for biological specimens. However, CO_2 is immiscible with water and should be exchanged with a solvent like ethanol-soluble in both water and liquid CO_2. On drying, biological specimens are usually coated with metal like gold, platinum, or palladium to make surfaces conductive electrically to characterize samples by SEM.

For TEM observations, more elaborate techniques include freeze-fracture electron microscopy [13]. This technique involves rapid sample cooling by placing it into liquid ethane or propane at the temperature of 160°C, the frozen droplet gets fractured with a cold knife and the fracture path follows the lowest resistance route and the contours of the nanostructures at the fracture plane. An electron transparent replica is made by evaporating a thin layer of carbon on the fracture plane. The contrast required to visualize the relief structure of the surface is created by evaporating a metal, such as platinum, on the carbon film at an angle of 45° (shadowing). CryoTEM and subsequent imaging by low-dose cryogenic TEM can also be envisioned to study aqueous, e.g., biological specimens in their near native hydrated state [12, 14, 15].

A microliter drop of the sample is added to a carrier grid and reduced in volume 5000 times by blotting with filter paper to get a thin film of a thickness of 100 nm having a large surface to volume ratio (SVR) which enables rapid heat exchange and vitrification efficiently. However, under ambient conditions, this SVR may result into rapid evaporation of the solvent accompanied by tremendous increase in the sample concentration and leading to a toppling of the embedded nanostructures.

3.20 ATOMIC FORCE MICROSCOPY (AFM)

Principle: The sample surface is scanned, by contacting it with a sharp tip at the edge of a microcantilever while the feedback electronics raises or lowers the tip, according to changes in surface topography, and to hold on to a constant position where the laser strikes the photodiode.

Characterization Techniques for Bio-Nanocomposites

3.20.1 THEORY

Atomic force microscopy (AFM) was introduced in the eighties along with another scanning probe microscopy (SPM) called scanning tunneling microscopy (STM) [16]. Unlike STM, AFM doesn't require conductive samples, and therefore has been in used in many scientific fields, specifically in the characterization of biological samples. The AFM diagnoses the surface by getting in touch with a sharp tip at the edge of a microcantilever Height and by reflecting a laser beam off the back surface of the cantilever onto a photodiode.

In the contact mode, the voltages are required to either rise or lower the tip serve as the height input for the image. Then, the set of displacement signals in the z-direction are taken as the topographic image of the specimen in question. However, if the sample's surface is soft enough amenable to deformation upon interaction of the sharp tip and the sample surface, the z-displacement signal no more shows the surface topography of the specimen under study. To get over these limitations, intermittent or noncontact modes of imaging or tapping modes have been employed. In tapping mode, the AFM cantilever is subjected to vibration at a given frequency close to its resonance frequency. The oscillation amplitude reduces in the neighborhood of the surface. Tapping mode operation takes advantage of amplitude signal for the purpose of feedback control. This mode operates to ensure a minimum contact in between the probe and the sample, for a very short span of time during a vibration cycle and as the tip contacts the surface intermittently, the tip deformation problem is checked, and therefore AFM is highly valuable tool to examine biological sample specimens and to have a glance at cell surfaces. Moreover, tapping mode decreases the lateral forces employed by the tip to the sample during scanning, thereby subsequently decreasing sample removal.

3.20.2 APPLICATIONS OF AFM

AFM has below-mentioned applications:

- AFM imaging doesn't ask for any staining or coating, making the direct imaging of biological specimens possible in physiologically conducive environments.
- AFM imaging has been in use for the examination of dynamic bioprocesses at the nanoscale (NS), e.g., dynamic studies of DNA in liquid environments [17, 19] and of live cell imaging [20].

- Given the tip dimensions and the spring constant of the cantilever, the sample stiffness can be calculated.
- AFM has been undertaken to explore the intra and intermolecular forces of bio-molecules.
- In biological or bio-inspired environments, force measurements by attaching an oligo-nucleotide to the cantilever and its complementary strand to a surface, forces required to rupture bonds between complementary oligo-nucleotides are measured [21, 22]. Forces between biotin and avidin [23, 24] or streptavidin [25] have been measured by AFM.
- AFM a tool of choice to investigate bionanocomposites structures and properties, e.g., characterization of the surface morphology of antibacterial bionanocomposite films made up of silk and zinc oxide nanoparticles [26] or in the analysis of the surface of hybrid platforms made up of silver nanoparticles and various kinds of biological molecules, e.g., chitosan, lipids, chlorophyll, and curcumin [27].
- AFM is also used to find out the mechanical properties of biofilms of chitin and silica at the NS [28].

3.21 SCATTERING TECHNIQUES

Principle: When a monochromatic, beam is made to interact with a sample, the intensity of the scattered radiation is measure of the scattering angle.

3.21.1 THEORY

The scattering vector (Q) is an important variable in the scattering experiments and the magnitude of Q depends upon the scattering angle and wavelength.

$$Q = \frac{4\pi \, Sin\,\theta/2}{\lambda}$$

The distance probed in scattering experiment is inversely proportional to Q ($\sim2/Q$), which means small Q are required for investigating large scale structures which can be accomplished with a right combination of high wavelength and low scattering angle. A wavelength that is equal to, or greater than, the particles size subjected to light scattering is generally used while for X-ray and neutron scattering, a low scattering angle is preferably used [29].

3.21.2 APPLICATIONS

Small Q refers to the scattering which is dependent upon the size and shape of the particles, while large Q it reflects upon the internal elucidation of structure of the particles, e.g., crystalline materials internal structure would give Bragg's peaks corresponding to typical planes of atoms and thereby therefore their unique lattice patterns.

3.22 DYNAMIC LIGHT SCATTERING (DLS)

Principle: Dynamic light scattering (DLS) or photon correlation spectroscopy (PCS) measures Brownian motion (i.e., random collision of particles dispersed within the solvent molecules) which is ascribed to the size of the particles, and that smaller particles move randomly and rapidly than coarser ones.

3.22.1 THEORY

DLS is based on the analysis of the variations in the intensity of the scattered light on illumination of particles. The size of a particle and its mobility because of Brownian motion is deduced by the Stokes-Einstein equation. A monochromatic radiation with frequency, 0, is incident then the colloid gives a spectrum having a Lorentzian shape, with the peak width which is determined from the product of diffusion coefficient of the particles, D, and Q^2:

$$I\left(\acute{E}\right) \infty \frac{DQ^2}{\left(\acute{E} - \acute{E}_0\right)^2 + \left(DQ^2\right)^2}$$

The hydrodynamic radius a is calculated from the diffusion coefficient, for a given solvent with viscosity, η.

$$a = \frac{kT}{6\pi\eta D}$$

3.23 ZETAMETRY

Principle: Zeta potential () that sets up at the interface within the diffuse layer inside which the ions and particles form a stable entity is measured.

3.23.1 THEORY

DLS is normally coupled with zetametry, a technique that provides for the determination of the electrostatic properties of colloids. Both kinds of analyses are done by using commercially available apparatus. The net charge developed at the surface of particle influences the way ions distribute in the neighboring interfacial region, leading to rise in concentration of counter ions (oppositely charged ions) in the immediate vicinity of the surface. Hence, an electrical double layer is found to form around each particle. The liquid layer around the particle has two parts: an inner one referred to as stern layer, at which the ions are firmly adhered and an outer one called diffuse layer where they are comparatively loosely held. Within the outer layer, the ions and particles form a stable entity inside the boundary. The potential that sets up at this boundary is referred to as the zeta potential, which is expressed in mV.

3.23.2 APPLICATIONS

The magnitude of the zeta potential measures the stability of the potential for given colloidal solution. If the particles in colloidal dispersion have a either too negative or positive zeta potential say + 30 mV, they have a tendency to repel one another, and don't aggregate and settle. Such colloidal systems are considered as stable. Moreover, the particles having low zeta potential values tend to agglomerate and flocculate. The key factor that influences zeta potential is pH. A zeta potential value without a mention of pH is meaningless. The pH value at which the zeta potential is zero is called the isoelectric point. It is the value at which the given colloidal solution has the lowest stability. One more influential factor affecting zetametry measurement is the ionic strength. The zeta potential depends upon the kind and strength of ions that interact with the particle surface in solution. Thus, it is of vital importance to conduct experiments using an electrolyte (say NaCl) at low concentration.

3.24 PARTICLE BIOMOLECULE INTERACTIONS INVESTIGATION

Particle biomolecule interactions investigation is an analytical challenge that asks for the detail study of effect of one another in isolation. In inorganic bioorganic interfacial systems, the soft system is vulnerable to structural modifications. The detection of such modifications can be done either

Characterization Techniques for Bio-Nanocomposites 73

directly or by their impact on the biomolecule reactivity. Such techniques find widespread use in fields like biochemistry and biology.

3.25 ELECTROPHORESIS

Principle: Electrophoresis technique is based on the separation of charged particles when an applied electric field (EF) is applied through polymer gel matrices normally agarose or polyacrylamide [30].

3.25.1 THEORY

When an EF of field strength E, is applied to a mixture of electrically charged biomolecules then they migrate to the electrode of opposite charge. However, different molecules migrate at different rates based on difference in their physical characteristics of the biomolecule and the polymer gel. Thus, the velocity of migration, of a charged molecule under the influence of EF is given as:

$$v = \frac{Eq}{f}$$

where, f – is the frictional coefficient; and q – is the net charge carried by the molecule.

The frictional coefficient refers to the resistance offered to mobility and is dependent upon the molecular mass, its compactness (molecular conformation), and buffer viscosity, other than the gel matrix porosity which allows the molecular migration [30]. This technique allows loading of molecules either in their native state, preserving their conformation and higher-order structure, or the sample may be added upon by a chemical denaturant which can give an unstructured linear chain whose mobility is governed by its length and mass to charge ratio.

Electrophoresis may be followed by the use of other techniques to unravel the separated species, e.g., ethidium bromide or silver staining.

3.25.2 APPLICATIONS

Electrophoresis is also a useful tool in colloidal science and can be applied:

- To separate complex mixtures with very high resolution. Two-dimensional gel electrophoresis (2-DE) is used to analyze proteins.

- To separate molecules linearly based on their isoelectric point in the first dimension and molecular mass in the second one [31].
- To isolate small inorganic clusters like gold [32] or semiconductors and magnetic particle clusters [33]).
- To separate the nanoparticles with a discrete number of functional groups [34], e.g., gold nanoparticles conjugated to nucleic acids [35, 37], or peptide conjugated semiconductors [38].

3.26 BIONANOCOMPOSITE PROPERTIES TESTING MATERIAL PROPERTIES

Bionanocomposites combine the properties of inorganic nanoparticles with those of biological ones. The NS dimensions of the inorganic material have a direct impact on the functionality of the biomaterial and hence the subsequent properties of the bionanocomposites such as optical, mechanical, and magnetic.

3.26.1 OPTICAL PROPERTIES

Optical properties are measured in terms of absorption and emission of light in the UV-Vis region of the electromagnetic (EM) spectrum. These techniques are commonly employed in a transmission mode for liquid and colloidal samples till the particle size is not too large as otherwise scattering effects may hinder the reading out of absorption bands at shorter wavelength and/or when the suspension is stable. For a solid material, scattering effects can be increased, and re-absorption of emitted light, may also take place. Further, use conventional sample holder made up of glass or quartz cuvettes is practically not possible. Equipment used for the analysis of solids is the integrating sphere that provides for collecting the light transmitted or reflected by a given sample to address the problems of conventional method.

3.26.2 MECHANICAL PROPERTIES

Mechanical properties encompass a wide range of properties that all reflect the structural response of a material when subjected to a mechanical force. Varied kinds of mechanical stresses can be applied to the sample of a material under study: compression, traction, bending, twisting, and so on. Mechanical measurements offer information either at its own scale or, at the

Characterization Techniques for Bio-Nanocomposites

scale based on application of local stress on the material. each method used for mechanical testing provides distinct and complementary information, and the choice of right tests depends upon the purpose for which the study is undertaken, keeping in view the stress(es) to which the material is to be applied in the application of interest. Mechanical testing lays emphasis upon the magnitude of force (termed stress) required to obtain a given structural response (termed strain) rather than on the response obtained by application of a given force.

3.26.2.1 MECHANICAL TESTING OF HYDROGELS

Hydrogels as an example that constitute a major chunk of bionanocomposites in bulk is examined. The mechanical properties of hydrogels depend largely upon the presence of water, e.g., under compression, water expulsion takes place, amounting to the measured stress. Assuming material porosity (i.e., water diffusion), this contribution can be calculated in theoretical terms, to derive the properties of the solid network only. Comprehensive analysis using either of these techniques is time taking. Under such circumstances, the material may dry out, accompanied by change in its mechanical behavior. Therefore, it is essential to retain high humidity in the environment to which the sample is kept exposed, such as by keeping it immersed in an aqueous solution.

Mechanical measurements can give insights into the structure of bionanocomposites and, in particular, how a rise in nanoparticle concentration within a biopolymer matrix can be advantageous for some mechanical properties and limiting others [45], e.g., when ZnO nanorods are embedded within a fish gelatin gel matrix. The samples are prepared as thick films and preserved at 50% relative humidity. The Young modulus and tensile strength (i.e., stress at break) increased with the nanoparticle content. Therefore, the materials are becoming stiffer and stronger. However, the elongation (i.e., strain) at break is decreasing with increasing ZnO content, that is, the materials become more brittle [46, 47].

3.26.2.2 DYNAMIC MECHANICAL ANALYSIS (DMA)

As in the rheological studies, the use of sinusoidal stress or strain can offer interesting information in reference to the contribution of the viscous behavior of the material. Dynamic mechanical analysis (DMA) is used in preference to study phase transitions (PTs), like the glass transition of polymers.

3.26.2.3 INDENTATION

Indentation techniques refer to the penetration of an indenter (probe) to form an indentation or cavity on a material specimen. They help in determining the hardness of a given material. Hardness is not measured in absolute terms but in comparison with the material making the probe which should necessarily be harder than the material under investigation. Most popularly, a diamond probe is used, say in the so-called Vickers hardness test. In this test, the surface of the indentation (imprint) left by the square-shaped tip of the probe onto the material surface as a function of the force applied by the indenter or probe helps in the calculation of the hardness index. Various modifications have been made in accordance with their adaptability to soft materials, small volumes, or TFs, e.g., nano-indentation tools which are designed, can be hyphenated with SEM or AFM for getting precise results about the imprint.

3.26.2.4 COMPRESSION TESTS

Compression tests are carried out by placing the sample on a fixed bottom plate and a vertically applying pressure with a mobile cylinder. The force (stress) required to reach a distance smaller than the initial thickness of the sample is recorded and plotted against the % change in sample thickness (strain).

Several important parameters can be obtained. The slope of the stress versus strain curve corresponds to the elastic modulus, called Young modulus (E). High E values indicate that a strong force is needed to obtain a substantial strain, when the material is stiff, against soft materials having low E values. The stress and strain at which the material breaks or becomes inelastic (end of the elastic domain) shows the yield strength of the material. The strain at breakpoint indicates its brittleness. While the extent of the plasticity reflects its malleability.

3.26.2.5 TENSILE TESTS

Tensile tests resemble with compression ones except that the sample is subjected to axial force in the former. Hence, the force necessary to increase the distance between the two ends of the material is recorded. Indeed, practically, it means that the sample must be tightly held by the two mobile pieces of the equipment, or grips. The resulting typical stress versus strain curves is

Characterization Techniques for Bio-Nanocomposites 77

similar to that obtained in compression tests: a linear part in the curve corresponds to the elastic domain follows either by rupture or by a plastic domain. In tensile tests, the extent of the plasticity reflects the material ductility.

3.26.2.6 RELAXATION TESTS

In these tests, the strain (i.e., the space between the plates or between the grips) is kept constant and the force exerted by the sample as a function of time is monitored. This allows depending upon the ability of the material to reorganize its internal structural to cope with its new dimensions and can give quite interesting insights about the structure, arrangement, and interactions of the molecules or particles within the material.

3.26.2.7 RHEOLOGY

Rheology can be defined as the study of the way in which matter flows in response to an applied shearing force.

Under such shearing stress, two extreme responses can be obtained:

- A viscous response (for liquids); and
- An elastic response (for solids).

However, soft materials, such as polymer-based solids, exhibit an intermediate viscoelastic behavior. A typical measurement involves a cell made of a fixed wall and a rotating wall. Three configurations exist: Couette cell, with two coaxial cylinders with the inner one being able to rotate, and plane plate (parallel plate) or cone plate, having a fixed bottom horizon plate and a rotating top horizontal plate or cone-shaped piece. The amenability of the geometry depends upon nature (liquid, particle suspension, solid) of the material to be examined. The sample has to be placed between the two walls, the mobile part is kept rotating at a fixed strain and the commensurate stress (i.e., the force necessary to keep the strain constant) is monitored and recorded.

3.27 MAGNETIC MEASUREMENTS

Different solids show a host of magnetic behaviors. However, unlike limited diversity in mechanical testing equipments, typical methods have been developed to get over the constraints in terms of field intensity or required sensitivity. Bionanocomposite materials can be analyzed by using two common techniques for their magnetic behaviors:

- Vibrating sample magnetometer (VSM); and
- Superconducting quantum interference device (SQUID).

Principle: The temporal variation in the intensity of a MF B produced by a given object induces the production of an EF in accordance with the Maxwell Faraday equation.

Therefore, if a magnetic sample is placed between conductive coils and vibrate, an electrical current gets generated and can be measured, whose intensity depends upon the MF intensity. The magnetic samples contain paramagnetic species, like, atoms, ions, clusters, or molecules containing unpaired electrons with electronic spin S, e.g., transition metals. Further, the magnetic interactions take place between the different species in the object, and placed in an external MF H by electromagnets. The magnetic behavior is dependent on the temperature. The simplest available equipment is VSM which is characterized by a good sensitivity, measurable at room temperature (RT). The most common equipment is a VSM with SQUID, wherein, the coils are superconducting, and the magnetic flux due to vibration of the magnetic sample can be detected via Josephson junctions. The resulting sensitivity is twice than that of VSMs, which allows for the study of single particles [48, 49].

3.28 BIOLOGICAL PROPERTIES

Incorporation of biomolecules within a material adds properties like biocatalysis, cyto/biocompatibility, or specific binding/recognition capabilities. However, if these properties demand distinct analytical protocols, then the relevant analytical tools (such UV-Vis spectrophotometry which can follow enzymatic activity, optical, FLU, and electron microscopy for cellular and tissue studies, ITC/SPR for binding affinities) are to be used.

There are a number of situations wherein normal biological tests can't be applied, or should be undertaken cautiously in presence of inorganic particles particularly when objects are in colloidal state as against isolated biomolecules, Bionanocomposite nanoparticles are vulnerable to aggregation. Moreover, they exhibit a large surface area, to be coated in a nonspecific manner by molecules present in their environment (e.g., proteins in culture medium).

Evaluation of these of interactions among bionanocomposite nanoparticles and cells is affected by:

- The state of aggregation and nature of the coating influence their interactions with cell membrane and internalization processes;
- Experimental conditions (ionic strength, temperature, concentration, culture medium composition) [50].

A second issue is related to the possible interference of the physical properties of the inorganic particles with the probe molecule or reaction. A recent survey of six common assays for protein quantification (BCA, Bradford), proliferation/viability (MTS/Alamar Blue), oxidative stress (catalase), and cytotoxicity (LDH) showed that nanoparticles do affect, especially at lower concentrations of say 0.1 mg.ml^1 [51].

3.29 CONCLUSION

Biopolymer NCs developed in the recent past have shown that the availability of the nanofiller such as graphene (GR) or clay increases the mechanical properties of polymer matrices to a great degree which exist in nanocomposite (solids, foams, and hydrogels). That is instrumental in contributing sound mechanical properties, such as high aspect ratio, surface area, mobility, and flaw tolerance of the nanofiller during testing, and/or for bettering the interfacial adhesion between the polymer and the nanofiller.

In the case of solid NCs, theories pertaining to composites such as Halpin Tsai and Mori Tanaka models can be employed to explore the elastic modulus volume fraction co-relation in biopolymer-based NCs, given that the effective volume fractions of nanofiller are used, which takes into account the contribution made by the polymer molecules (surface-immobilized) to further the reinforcing effect. In the case of porous NCs, the nanofiller modify the porous structure and enhance the specific strength and modulus of polymer foams. The relative modulus and density relationships of low-density biopolymer nanocomposite foams (open or closed cell), can be determined by employing the normalized Gibson Ashby model. In the case of nanocomposite (hydrogels), the presence of nanofiller can not only enhance the mechanical properties of it but also work as a physical crosslinking agent to form a hydrogel wherever possible.

Due to the outstanding mechanical properties, biopolymer-based NCs can be investigated as potential candidates for a variety of biomedical applications including medical devices, tissue scaffolds, artificial tissues and organs, drug delivery systems, wound dressings, surgical tools, biosensors, etc.

KEYWORDS

- **atomic force microscopy**
- **auger electron spectroscopy**
- **biomolecules**
- **bio-nanocomposites**
- **dynamic light scattering**
- **inorganic-biological interface**

REFERENCES

1. Montano, M. D., Olesik, J. W., Barber, A. G., Challis, K., & Ranville, J. S., (2016). *Anal. Bioanal. Chem., 408*, 5053–5074.
2. Contado, C., & Pagnoni, A., (2008). *Anal. Chem., 80*, 7594–7608.
3. Muller, C. M., Pejcic, B., Esteban, L., Delle, P. C., Raven, M., & Mizaikoff, B., (2014). *Sci. Rep., 4*, 6764.
4. Turner, N. H., Dunlap, B. I., & Colton, R. J., (1984). *Anal. Chem., 56*, 373R–416R.
5. Baer, D. R., & Engelhardt, M. H. J., (2010). *Electron Spectrosc. Relat. Phenom., 178, 179*, 415–432.
6. Shavorskiy, A., Aksoy, F., Grass, M. E., Liu, Z., Bluhm, H., & Held, G., (2011). *J. Am. Chem. Soc., 133*, 6659–6667.
7. Wang, X., Masse, S., Laurent, G., Helary, C., & Coradin, T., (2015). *Langmuir, 31*, 11078–11085.
8. Energy Dispersive Analysis, (2015). *Essential Knowledge Briefings* (2nd edn.). John Wiley & Sons, Ltd, Chichester.
9. Egerton, R. F., (2009). *Rep. Prog. Phys., 72*, 016502.
10. Arenal, R., De Matteis, L., Custardoy, L., Mayoral, A., Tence, M., Grazu, V., De La Fuentes, J. M., Marquina, C., & Ibarra, M. R., (2013). *ACS Nano, 7*, 4006–4013.
11. Davis, S., (2010). Electron microscopy. In: Cosgrove, T., (ed.), *Colloid Science: Principles, Methods and Applications* (2nd edn.). John Wiley & Sons, Ltd, Chichester.
12. Friedrich, H., Frederik, P. M., De With, G., & Sommerdijk, N. A. J. M., (2010). *Angew. Chem. Int. Ed., 49*, 7850–7858.
13. Mondain-Monval, O., (2005). *Curr. Opin. Colloid Interface Sci., 10*, 250–255.
14. Parry, A. L., Bomans, P. H. H., Holder, S. J., Sommerdijk, N. A. J. M., & Biagini, S. C. G., (2008). *Angew. Chem. Int. Ed., 47*, 8859–8862.

Characterization Techniques for Bio-Nanocomposites

15. Newcomb, C. J., Moyer, T. J., Lee, S. S., & Stupp, S. I., (2012). *Curr. Opin. Colloid Interface Sci., 17*, 350–359.
16. Binnig, G., Quate, C. F., & Gerber, C., (1986). *Phys. Rev. Lett., 56*, 930–933.
17. Kasas, S., Thomson, N. H., Smith, B. L., Hansma, H. G., Zhu, X., Guthold, M., Bustamante, C., et al., (1997). *Biochemistry, 36*, 461–468.
18. Radmacher, M., Fritz, M., Hansma, H. G., & Hansma, P. K., (1994). *Science, 265*, 1577–1579.
19. Wang, H., Bash, R., Yodh, J. G., Hager, G., Lindsay, S. M., & Lohr, D., (2004). *Biophys. J., 87*, 1964–1971.
20. Allison, D. P., Mortensen, N. P., Sullivan, C. J., & Doktycz, M. J., (2010). *Wiley Interdiscip. Rev. Nanomed. Nanoiotechnol., 6*, 618–634.
21. Lee, G. U., Chrisey, L. A., & Colton, R. J., (1994). *Science, 266*, 771–773.
22. Boland, T., & Ratner, B. D., (1995). *Proc. Natl. Acad. Sci. U.S.A., 92*, 5297–5301.
23. Florin, E. L., Moy, V. T., & Gaub, H. E., (1994). *Science, 264*, 415–417.
24. Moy, V. T., Florin, E. L., & Gaub, H. E., (1994). *Science, 266*, 257–259.
25. Lee, G. U., Kidwell, D. A., & Colton, R. J., (1994). *Langmuir, 10*, 354–357.
26. Kiro, A., Bajpai, J., & Bajpai, A. K., (2016). *J. Mech. Behav. Biomed. Mater., 65*, 281–294.
27. Barbinta-Patrascu, M. E., Badea, N., Pirvu, C., Bacalum, M., Ungureanu, C., Nadejde, P. L., Ion, C., & Rau, I., (2016). *Mater. Sci. Eng. C Mater. Biol. Appl., 69*, 922–932.
28. Smolyakov, G., Pruvost, S., Cardoso, L., Alonso, B., Belamie, E., & Duchet-Rumeau, J., (2016). *Carbohydr. Polym., 151*, 373–380.
29. Richardson, R., (2010). Scattering and reflection techniques. In: Cosgrove, T., (ed.), *Colloid Science. Principles, Methods and Applications* (2nd edn.). John Wiley & Sons, Ltd, Chichester.
30. Magdeldin, S., Zhang, Y., Bo, X., Yoshida, Y., & Yamamoto, T., (2012). *Biochemistry, Genetics and Molecular Biology. Gel Electrophoresis: Principles and Basics*. InTech., Rijeka, Croatia.
31. Magdeldin, S., Enany, S., Yoshida, Y., Xu, B., Zhang, Y., Zureena, Z., Lokamani, I., Yaoita, E., & Yamamoto, T., (2014). *Clin. Proteomics, 11*, 16.
32. Schaaff, T. G., Knight, G., Shafigullin, M. N., Borkman, R. F., & Whetten, R. L., (1998). *J. Phys. Chem. B, 52*, 10643–10646.
33. Roullier, V., Grasset, F., Boulmedais, F., Artzner, F., Cador, O., & Marchi, A. V., (2008). *Chem. Mater., 20*, 6657–6665.
34. Sperling, R. A., Pellegrino, T., Li, J. K., Chang, W. H., & Parak, W. J., (2006). *Adv. Funct. Mater., 16*, 943–948.
35. Alivisatos, A. P., Johnsson, K. P., Peng, X., Wilson, T. E., Loweth, C. J., Bruchez, Jr. M. P., & Schultz, P. G., (1996). *Nature, 382*, 609–611.
36. Zanchet, D., Micheel, C. M., Parak, W. J., Gerion, D., & Alivisatos, A. P., (2001). *Nano Lett., 1*, 32–35.
37. Ackerson, C. J., Sykes, M. T., & Kornberg, R. D., (2005). *Proc. Natl. Acad. Sci. U.S.A., 102*, 13383–13385.
38. Dif, A., Henry, E., Artzner, F., Baudy-Floch, M., Schmutz, M., Dahan, M., & Marchi-Artzner, V., (2008). *J. Am. Chem. Soc., 130*, 8289–8296.
39. Billstein, P., Wahlgren, M., Arnebrant, T., McGuire, J., & Elwing, H., (1995). *J. Colloid Interface Sci., 175*, 77–82.
40. Freyer, M. W., & Lewis, E. A., (2008). *Methods Cell Biol., 84*, 79–113.

41. Falconer, R. J., & Collins, B. M., (2011). *J. Mol. Recognit., 24*, 1–16.
42. Limo, M. J., & Perry, C. C., (2015). *Langmuir, 31*, 6814–6822.
43. Aime, C., Mosser, G., Pembouong, G., Bouteiller, L., & Coradin, T., (2012). *Nanoscale, 4*, 7127–7134.
44. Cedervall, T., Lynch, I., Lindman, S., Berggard, T., Thulin, E., Nilsson, H., Dawson, K. A., & Linse, S., (2007). *Proc. Natl. Acad. Sci. U.S.A., 104*, 2050–2055.
45. Rouhi, J., Mahmud, S., Naderi, N., Ooi, C. H. R., & Mahmood, M. R., (2013). *Nanoscale Res. Lett., 8*, 364.
46. Asthana, A., Momeni, K., Prasad, A., Yap, Y. K., & Yassar, R. S., (2011). *Nanotechnology, 22*, 265712.
47. Casillas, G., Palomares-Baez, J. P., Rodriguez, L. J. L., Luo, J., Ponce, A., Esparza, R., Velazquez-Salazar, J. J., Hurtado-Macias, A., Gonzalez-Hernandez, J., & Jose, Y. M., (2012). *Philos. Mag., 92*, 4437–4453.
48. Balanda, M., (2013). *Acta Phys. Pol. A, 124*, 964–976.
49. Naik, R., Senaratne, U., Powell, N., Buc, E. C., Tsoi, G. M., Naik, V. M., Vaishnava, P. P., & Wenger, L. E., (2005). *J. Appl. Phys., 97*, 10J313.1–10J313.3.
50. Del Pino, P., Pelaz, B., Zhang, Q., Maffre, P., Nienhaus, G. U., & Parak, W. J., (2014). *Mater. Horiz., 1*, 301–313.
51. Ong, K. J., MacCormack, T. J., Clark, R. J., Ede, J. D., Ortega, V. A., Felix, L. C., Dang, M. K. M., et al., (2014). *PLoS One, 9*, e90650.

CHAPTER 4

Chemistry Matrices of Biotransformation

SHRIKAANT KULKARNI

Assistant Professor, Vishwakarma Institute of Technology, Department of Chemical Engineering, 666, Upper Indira Nagar, Bibwewadi, Pune – 411037, India, Mobile: +91-9970663353, E-mail: shrikaant.kulkarni@vit.edu

ABSTRACT

Green and sustainable chemical manufacturing can be done using suitable catalysts as well as by employing meaningful green matrices for ascertaining the greenness of the processes and quantitation of reaction efficiency. Green chemistry matrices will also help in monitoring the kinetics of the reactions, e.g., atom efficiency and E factors are very widely recognized matrices across the board. If the green chemistry matrices are used in conjunction with life cycle assessment (LCA), it will provide sound and concrete parameters for assessment of greenness and sustainability of spectrum of chemical processes. Catalysis like homogeneous, heterogeneous, organo-catalysis and bio-catalysis will play an important role in order to evolve at sustainable technologies. Bio-catalysis, namely, has a comparative advantage in this regard in terms of mild reaction conditions with a catalyst which is biocompatible, biodegradable, and renewable with step economic, and highly selective processes reflecting upon better product quality, economical, and reduced waste generation. Moreover, the other benefits accrued include process intensification by virtue of integration of bio-catalytic steps with cascading effect in processes. Bio-catalysis, will further the transition from a global economy relying upon non-renewable fossil feedstocks to a sustainable one which is renewable resources based driven by greening of the synthetic routes or pathways.

This chapter highlights a discussion of host of green chemistry matrices that are put into use for ensuring the greenness and sustainability in the manufacturing processes.

4.1 INTRODUCTION

The organic synthesis dated back to the Wohler's synthesis of urea as the natural product produced from ammonium isocyanate as a feedstock (1828). This puts an end to the vis vitalis (vital force) theory, according to which a substance produced as a result of the metabolic activity of a living organism is not so easy to be produced synthetically. This discovery had huge impact as it prepared the groundwork, for the synthesis of various organic chemicals in the laboratory. Another achievement of vital importance was the synthesis of the synthetic dye, mauveine (aniline purple), on mass scale [1]. The preparation of the antimalarial drug, quinine, using N-allyl toluidine oxidation with potassium dichromate. This discovery paved the way to lay foundation for the synthetic dyestuffs industry using coal tar, a waste byproduct generated during steel manufacturing. The synthesis of mauveine was followed by the natural dyes alizarin and indigo. The production of such dyes at commercial scale has led to the stoppage of their agricultural production and has given birth to a science-based, followed by a German-based, chemical industry. The pharmaceutical and allied fine chemical industries of present-day are nothing but spin-offs to dyes industry. Synthetically developed molecules were found to be simple in the beginning that turned into complicated one, over the years, e.g., the regio- and enantiospecific chemical synthesis involving microbial hydroxylation of progesterone to 11α-hydroxyprogesterone was like a landmark synthesis [2] which further led to the feasible and economically viable synthesis of cortisone and in turn of steroid hormones at commercial level. It marked the development of drugs accompanied by continuously increasing molecular complexity, and through untiring R&D efforts chemists have been working to meet the ever growing needs of sophistication, e.g., the anticancer drug, taxol [3] as a result of its semisynthesis from baccatin III, extracted from the English yew, Taxus baccata. Therefore, the success of the modern pharmaceutical industry has taken place as a result of the giant work undertaken over the decades in the organic synthesis. A great many number of synthetic methods which time tested and used widely were developed when the hazardous nature of many chemicals was

Chemistry Matrices of Biotransformation

not yet ascertained and waste minimization as well as sustainability were not the issues of much concern.

However, over the last two decades, much attention has been paid to the fact that the those classical synthetic methodologies are encouraged which would either eliminate or minimize the waste so as to be atom efficient, e.g., the synthesis of fluoroglucinol, a pharmaceutical intermediate [4] was produced till mid-1980s, mainly from 2,4,6-trinitrotoluene (TNT), an ideal example of vintage 19th-century organic chemistry which would have caught the attention of Perkin as shown in Figure 4.1.

$$C_7H_5N_3O_6 + K_2Cr_2O_7 + 5H_2SO_4 + 9Fe + 21NCl \rightarrow C_6H_6O_3 + CR_2(SO_4)_3 + 2KHSO_6 - 9FeCl_2 - {}_3NH_4Cl + CO_2 + OH_2$$

O ....(1)

Molecular weight of the parent (desired) product, $C_6H_6O_3 = 126$

Sum of molecular weight of all products $= 2282$

% Atom economy $= \dfrac{126}{2292}$ X 100 $= 5.5$

FIGURE 4.1 Atom economy calculation in the manufacture of fluoroglucinol.

This process produces fluoroglucinol with the yield >90% over the three reaction steps and, based on the classical concepts of selectivity and reaction efficiency, would normally be adjudged as a selective and efficient process. However, every kilogram of fluoroglucinol, produces nearly 40 kg of solid waste, containing mixture of salts like $Cr_2(SO_4)_3$, NH_4Cl, $FeCl_2$, and $KHSO_4$. It was subsequently discarded because of the heavy costs incurred for the disposal of waste laced with chromium. Although the reaction stoichiometry predicts 100% chemical yield and exact stoichiometric quantities of the various reagents still actually, the oxidant and reductant, large volume of sulfuric acid in excess, the later requires to be neutralized with base consequently, and therefore the yield of phloroglucinol practically is less than 100%.

The wasteful use of raw materials has led to the evolution of the E factor for evaluating the environmental impact of chemical manufacturing processes which demand a paradigm shift from conventional concepts of reaction efficiency and selectivity that focus preferentially on yield, rather than maximum utilization of raw materials, waste, and the use of toxic and/ or hazardous substances elimination. Synthetic organic chemists are well versed with terms like reaction selectivity (yield/conversion) and chemo-, regio-, stereo-, and enantioselectivity but unaware about the terms such as atom selectivity of a reaction before the mid-1990s, and is defined as the mass of the reactants actually used in the making of the product, while the leftover other than product is waste. This selectivity is of great significance from the point of view of environmental impact which is instrumental in the evolution of the concept of atom economy (AE).

4.2 GREEN CHEMISTRY FOR SUSTAINABILITY

The World Commission for Environment and Development came into existence in 1983 by the United Nations and was entrusted with the responsibility of preparing a report on the long-term, sustainable, and environmentally friendly development perspectives across the globe by 2000 and thereafter. This has resulted into a published document, Our Common Future some four years later [5], which is recognized as the Brundtland Report. The report defined the premises of the concept of sustainable development and in the following two decades, the sustainability concept has dragged the attention of both industry and the society on the whole. The term "Green Chemistry" was coined in the early part of 1990s by Paul Anastas [6] of the watchdog agency and the apex body of the US, Environmental Protection Agency (EPA). The EPA officially adopted the name "US Green Chemistry Program," which was instrumental in kick-starting the activities in the United States, like the Presidential Green Chemistry Challenge Awards and the annual Green Chemistry and Engineering Conference. However, it doesn't mean that research on green chemistry was nonexistent before 1990s; but, it did not have that necessary recognition. The major concept which guides green chemistry is benign by design of both products and processes [7]. This concept is encompassed in the 12 Principles of Green Chemistry [6], which in a nutshell are given as:

- Prevention of waste rather than remediation;
- Atom efficiency or economy;
- Use of less hazardous chemicals;
- Design for safer products;
- Innocuous solvents and auxiliaries;
- Design for energy efficiency;
- Use of renewable raw materials preferentially;
- Design shorter synthetic routes or pathways (avoid derivatization/ protection/auxiliary agents);

Chemistry Matrices of Biotransformation

- Use catalytic over stoichiometric reagents;
- Design products amenable to biodegradation;
- Safer and real time analytical methods;
- Design inherently safer processes.

Anastas and Zimmerman [8] proposed the 12 principles of green engineering, with the same underlying philosophy such as conserve energy and resources and avoid waste and hazardous materials as those of green chemistry, but with an engineering perspective [9]. Alternatively, the premises of the working definition of green chemistry can be given in a single sentence [4].

Green chemistry efficiently or economically utilizes (preferentially renewable) raw materials, eliminates or reduces the waste and avoids the use of toxic and/or hazardous substances and solvents in the synthesis and use of chemical products.

Raw materials refer to kind of energy source used in the process. Green chemistry provides for the elimination of waste at its source, which implies that pollution prevention is better than cure or waste remediation in the later stage after it is generated or looking for end-of-pipe solutions, as stated in the first principle of green chemistry. In the last 15 years, the concept of green chemistry has been gaining huge momentum and widely embraced by both industrial and academic communities equally. Sustainability is our ultimate but common goal and green chemistry can provide ways and means to accomplish it. Green Chemistry allows the comparison between the processes (and products) based on their extent of greenness. As nothing is perfect, in a same tone, we can say that there is nothing like absolute greenness; one process is greener than another in relative terms, as the beauty, or greenness depends on how one perceives it. Notwithstanding this, as Lord Kelvin said, "to measure is to know," and extent of greenness can be quantified by using green metrics and are a prerequisite for furthering the green chemistry knowledge as well as to meaningfully compare the greenness of processes.

4.3 GREEN CHEMISTRY MATRICES

4.3.1 ATOM ECONOMY (AE) AND THE ENVIRONMENTAL IMPACT FACTOR (E FACTOR)

It is now widely accepted that two key measures which quantify the potential environmental impact of chemical processes are the E factor [10–15], which is defined as the ratio of the mass of waste to the desired product and the atom economy, or atom efficiency (AE) introduced by Trost in 1991 [16], defined as the ratio of the molecular weight of the desired product to the sum of the molecular weights of all the products generated in the given chemical equation [17]. A thorough understanding of the equation showing stoichiometric amounts of reactants and products helps in predicting the theoretical amount of waste that is expected to be generated without undertaking experiments. Atom utilization helps in quickly assessing the environmental acceptability of alternative processes to a particular chemical [18–20]. In Figure 4.2, we compare the AE for the conventional synthesis of chlorohydrin pathway to propylene oxide with that of oxidation with the green oxidant hydrogen peroxide, where the byproduct is water [21].

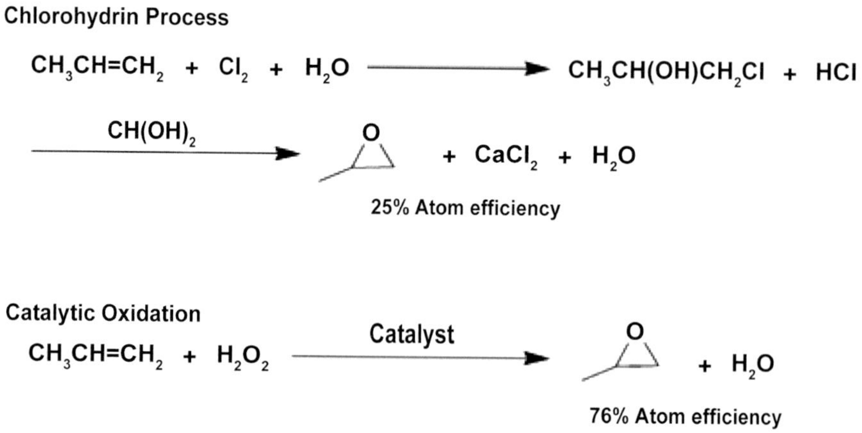

Chlorohydrin Process

$$CH_3CH=CH_2 + Cl_2 + H_2O \longrightarrow CH_3CH(OH)CH_2Cl + HCl$$

$$\xrightarrow{CH(OH)_2} \triangle^O + CaCl_2 + H_2O$$

25% Atom efficiency

Catalytic Oxidation

$$CH_3CH=CH_2 + H_2O_2 \xrightarrow{Catalyst} \triangle^O + H_2O$$

76% Atom efficiency

FIGURE 4.2 Atomic efficiency of processes for the manufacture of propylene oxide.

AE is considered as a theoretical number calculated based on how much part of reactants is used in the making of products and reactants are used in exactly stoichiometric amounts and it doesn't take into account host of other substances, such as solvent and acids or bases used in process, which don't figure in the stoichiometric equation. While, the E factor, is the actual quantity of waste generated in the process, which implies everything but the desired product. Unlike for AE calculation E factor considers the chemical yield and auxiliary reagents, solvent losses, as well as even the energy required (difficult to quantitate though) except water from the calculation as the E factor may rise to exceptionally high value in many cases and thereby making comparison of processes difficult meaningfully, e.g., for an aqueous waste stream, only the inorganic salts and organic compounds contained in

Chemistry Matrices of Biotransformation

the water are considered, excluding water. However, in the pharmaceutical industry, water is very much included in the calculation of E factor and water usage is a vital issue particularly in biomass conversion and fermentation processes.

A high E factor represents larger waste, and subsequently, more negative environmental impact. The ideal E factor is zero. In simple terms, it is ratio of kilograms (of feedstock) in, subtracted by kilograms of desired product, to kilograms of the product produced. The calculation of it is very simple based on the knowledge of the amount of starting materials (say tons) purchased and that of the product sold, for a given product produced by a whole company. Water used in the process is found to be excluding but not water generated. The quantum of magnitude of the waste problem is visible from E factors in different segments of the chemical industry as shown in Table 4.1. It shows that the E factor increases substantially from sectors bulk chemicals to fine chemicals and pharmaceuticals.

TABLE 4.1 E Factor

Industry	Volume (tons/annum)	E Factor Kg Waste/Kg Product)
Pharmaceutical industry	$10–10^3$	25 to > 100
Fine chemicals	$10^2–10^4$	5 to > 50
Bulk chemicals	$10^4–10^6$	< 1–5

On the one hand, the target compounds which at times are more complex molecules as against bulk chemicals and their synthetic routes involve multiple steps, which are amenable to produce large amount of waste. There the use of stoichiometric reagents in these industry segments is encouraged. While in bulk manufacture of chemicals, the production volumes are huge and the use of many stoichiometric reagents due to economic constraints is normally restricted.

The E factor has been widespread in its adoption by the chemical industry and, in preference, by the pharmaceutical industry [22] as a meaningful parameter for the environmental impact assessment of the manufacturing processes [23, 24]. The Green Chemistry Institute Pharmaceutical Round Table has an inventory of the data pertaining to the waste generated in processes to drive the greening of the many leading pharmaceutical companies namely, Eli Lilly, Glaxo Smith Kline (GSK), Pfizer, Merck, Astra-Zeneca, Schering Plough, and Johnson & Johnson. Other metrics too have been proposed for quantifying the environmental acceptability of processes (Figure 4.3) [25–27].

Other matrices can be classified into the following two types:

- Those based on the stoichiometric equation of the reaction just as a refinement of the AE concept; and
- Those which address the actual quantity of waste formed in the process thereby variations in E factor.

Constable et al. [28] at GSK introduced two concepts like reaction mass efficiency (RME) and carbon efficiency (CE). RME is defined as the ratio of mass of product produced to the total mass of the reactants in the stoichiometric equation expressed in percentage terms. It is considered as a refinement of AE which takes the cognizance of the yield of the desired product and the actual amounts of the reactants used. However, a limitation of RME as against AE is that it demands experimentation being done in the first place for the quick analysis of different processes to be done. CE resembles with RME but considers only carbon, that is, it is the ratio of mass of carbon in the product to the total mass of carbon in the reactants.

The use of mass intensity (MI), which is defined as the ratio of the total mass of materials used in a process to the mass of the product expressed in percentage, that is:

$$MI = E \text{ factor} + 1$$

with the ideal MI = 1 and the E factor = 0 [29].

The mass productivity (MP) is the reciprocal of the MI. An analogous metric: the effective mass yield (EMY) is proposed which is defined as the ratio of mass of the desired product to the total mass of the non-benign or unfriendly reactants used in its making. It does not take into account the environmentally friendly compounds, like NaCl and acetic acid which is questionable as these so considered environmentally benign compounds can have impact based on their volume. Further, defining the term as nonbenign is not so easy and arbitrary [28], therefore, the term EMY has ambiguity in its definition.

It means that none of these metrics has any comparative advantage over AE and the E factor for evaluating how is a process wasteful. The former is a quick tool that can be made use of before undertaking any experimental work and the latter measures the total waste that is actually generated in practice. Thus, AE and the E factor are the green metrics which complement one another. From the stoichiometry of overall phloroglucinol process as shown in Figure 4.4, the AE calculated is about 5%

$$E = \frac{\textit{Total mass of waste}}{\textit{Mass of target product}}$$

%Atom Efficiency (AE)

$$E = \frac{\textit{Molecular mass of product X 100}}{\textit{Sum of molecular mass of reactants}}$$

Mass Intensity (MI)

$$E = \frac{\textit{Total mass used in a process}}{\textit{Mass of product}}$$

$$E = \frac{\textit{Total mass of product or waste}}{\textit{Total Mass of reactants}}$$

% Reaction Mass Efficiency (RME)

$$E = \frac{\textit{Mass of product C X 100}}{\textit{Mass of reactant A + Mass of reactant B}}$$

Mass Productivity (MP)

$$E = \frac{\textit{Mass of product}}{\textit{Total Mass used in the process}}$$

% Carbon Efficiency (CE)

$$E = \frac{\textit{Amount of Carbon in product X 100}}{\textit{Total Carbon present in reactants}}$$

% Effective Mass Yield (EMY)

$$E = \frac{\textit{Mass of product X 100}}{\textit{Mass of hazardous reagents}}$$

FIGURE 4.3   Green matrices.

92 Green Chemistry and Green Engineering

and an E factor about 20 but in practice, the E factor is 40. This is attributed to yield less than 100% and a molar excess of the various reactants and the sulfuric acid used in large excess which requires neutralization with adequate amount of alkali in the experimental work. The high E factors for the processes used in the manufacture of fine chemicals and pharmaceuticals, and even some bulk chemicals, comprise of inorganic salts like NaCl, Na_2SO_4, and $(NH_4)_2SO_4$ obtained in the reaction or during neutralization steps.

4.4 THE NATURE OF WASTE

All the metrics so far take into account only the mass of the waste produced in a process. However, the environmental impact assessment of the waste generated should be done not only based upon its quantity but also its nature. E.g., One Kg of NaCl can't be held equivalent to 1 kg of a Cr salt. Hence, the term "environmental quotient," EQ, has been introduced [13] which is the product of the E factor and the unfriendliness quotient, Q assigned arbitrarily. E.g., Q value of 1 is assigned to NaCl and, 100–1000 to a heavy metal (chromium) salt, given its toxicity, ease of recycling, etc. The magnitude of Q is of course questionable and not so easy to measure but the important thing is that quantification of the environmental impact of chemical processes may, in principle, be possible [31]. It is to be kept in mind that Q for a given substance is affected by both the production volume and the production site location, e.g., the production of 100–1000 tons/annum of NaCl is not expected to pose any threat due to waste problem but 10,000 tons/annum, may do so due to disposal problem, thus there is bound to be an increase in Q. However, the Q value could decrease again as recycling by electrolysis is a viable option in the event of large quantities of NaCl generation. Thus, the Q value of a given waste can be estimated from its ease of disposal or recycling. Organic waste is, comparatively easy to dispose of than inorganic one. This is of particular importance whenever the green metrics for biocatalytic processes are considered. Another approach in evaluation of the environmental impact and subsequently the sustainability of both products and processes, is Life Cycle Assessment (LCA) in general [32, 33] in terms of quantitable environmental impact indicators, like energy usage, global warming, ozone depletion, acidification, eutrophication, smog formation, and ecotoxicity, apart from the waste generated.

4.5 WASTE MINIMIZATION USING CATALYSIS AND SUSTAINABLE REACTION MEDIA

The waste generated in the manufacture of fine chemicals and pharmaceuticals is primarily due to the use of stoichiometric inorganic and organic reagents that are either not used or partially used in the product formation. A cursory look at the phloroglucinol process shows peculiar examples: oxidations with chromium (VI) reagents and other inorganic oxidizing agents such as permanganate and manganese dioxide, and reductions with metals such as Na, Mg, Zn, and Fe and metal hydrides ($LiAlH_4$ and $NaBH_4$). Similarly, a host of reactions, for example, sulfonations, nitrations, halogenations, diazotizations, and Friedel-Crafts acylations, using stoichiometric quantities of mineral acids such as H_2SO_4, HF, H_3PO_4 and Lewis acids ($AlCl_3$, $ZnCl_2$, BF_3) generate waste. The solution to this problem is the substitution of stoichiometric methods with more clean catalytic options [11–15, 34], e.g., catalytic hydrogenation, oxidation, and carbonylation (Figure 4.4) are highly atom-economical, low-salt processes.

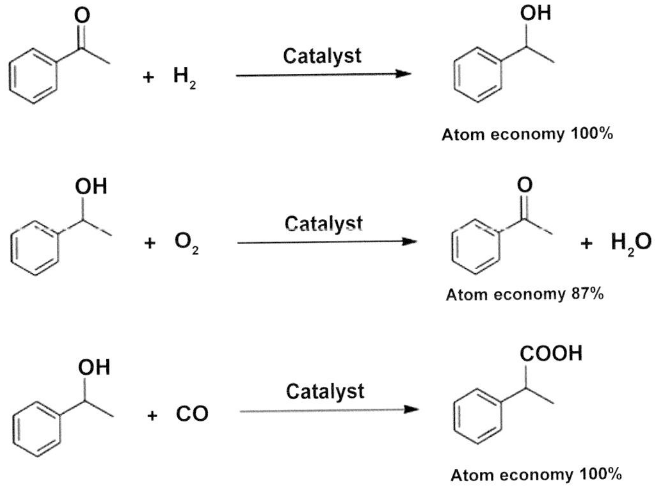

FIGURE 4.4   Atom-economical catalytic processes.

The problem of generation of huge amounts of inorganic salts can similarly be overcome by substituting mineral acids, such as H_2SO_4, Lewis acids, and bases, like NaOH and KOH, with recyclable solid acids and bases, preferably in catalytic amounts [35]. Recyclable solid catalysts are instrumental in minimizing waste in industrial organic synthesis [36]. Another major cause of waste is solvent losses, which normally is led to atmosphere or leach into ground water. Solvent losses are therefore responsible for contributing to the high E factors of processes of pharmaceutical industries [37]. Health and/or safety issues associated with several conventional organic solvents, like chlorinated hydrocarbons (HCs), have made their use being cut down substantially. Hence, pharmaceutical industries are emphasizing upon checking solvent use and evolving at alternatives which are more environmentally friendly to several conventional organic solvents, say chlorinated and aromatic HCs. If details are not known, about whether solvents would be recycled by distillation or otherwise then it is assumed that it would involve a 10% solvent loss.

Solventless processes are preferred most but if the use of solvent is indispensable, then it should be safe to use and should be amenable to efficiently remove it from the product and reuse. Several nonconventional and reaction media or designer solvents have been in use in recent years, including water [38], supercritical CO_2 [39], fluorous biphasic [40], and ionic liquids [41], in isolation or in liquid-liquid biphasic media. The use of water and supercritical carbon dioxide as reaction media caters to the current needs of using renewable raw materials, which are obtained from carbon dioxide and water.

As explained earlier, the reasonably high E factors in pharmaceuticals manufacturing processes can also be attributed to the multistep syntheses that are widely used in this industry segment which has led Wender [42] to promote the adoption of multistep economic syntheses. It can be accomplished by substituting conventional multistep syntheses with shorter catalytic initiatives. The ultimate goal is the adoption of catalytic processes which have a cascading effect, leading to integration of many catalytic steps with step-economical, one-pot procedures with no need to isolate the intermediates [43]. This is truly emulation or the elegant orchestration of enzymatic catalyzed steps followed in metabolic pathways of a living cell.

The advantages of shifting from multistep syntheses to catalytic cascades are:

- Less unit operations;
- Less solvent;
- Low reactor volume;
- Shorter cycle times;
- Higher volumetric and space-time yields;
- Less waste (lower E factor) which furthers economic and environmental benefits to a substantial extent;

Chemistry Matrices of Biotransformation

- Reactions can be coupled to drive equilibria toward the product, thus reducing the need for excess reagents.
- On the other hand, there are problems to be overcome:
- Catalysts are often not compatible (e.g., a biocatalyst like enzyme and a Chemical one like a metal);
- Rates and optimum conditions can differ a lot;
- Difficulty in recovery and recycle of catalyst.

Nature comes to our rescue in solving the problem of compatibility by segmentation of enzymes in various parts of the cell, which prescribes that the key in the development catalytic cascades is immobilization of the hosts of catalysts.

In a nutshell, waste generation can be ascribed to the use of stoichiometric reagents, multistep syntheses, and huge solvent losses, the waste problem can be addressed by substitution of stoichiometric reagents with atom efficient and step-economic catalytic procedures in renewable reaction media [44–46], which is achievable through true elegance and efficiency in organic synthesis.

4.6 BIOCATALYSIS

Biocatalysis has a multifaceted role to play in conformance with the principles of green chemistry such as:

- Reactions can be carried out under milder conditions at say physiological pH and ambient temperature and pressure using a biodegradable catalyst such as an enzyme which is derived from renewable materials and in an ecofriendly solvent like water.
- Reactions of multifunctional molecules take place with high degree of activity and chemo-, regio-, and stereoselectivity without activation of functional groups, protection, and deprotection steps required unlike conventional organic syntheses.
- Bring about processes which are multi-step but economical, less waste-producing, and are subsquently, both environmentally and economically more preferable than traditional pathways.
- Yield highly qualitative products against conventional chemical or chemocatalytic processes, e.g., problem of contamination with traces of (noble) metals is avoided, an issue of concern particularly in the manufacture of pharmaceuticals.

96   Green Chemistry and Green Engineering

- Enzyme catalyzed processes (except fermentations) can be carried out in standard multipurpose batch reactors and therefore doesn't ask for any additional investment.
- Advances in biotechnology have, offered the use of biocatalysis in industrial organic synthesis.
- Recombinant DNA techniques have offered to produce virtually any enzyme at affordable price for its commercial use.
- Modern protein engineering tools, like *in-vitro* evolution [47], have made it possible to manipulate the enzymes in order to exhibit the requisite substrate specificity, activity, stability, pH profile, etc.
- Optimize the enzyme to suit the needs of the predefined optimum process, which is, truely benign by design.
- The development of effective immobilization techniques has made possible the optimization of the storage, operational stability, recovery, and recycling of enzymes [48].
- Biocatalytic processes are carried out under almost the same set of conditions at ambient temperature and pressure, it allows for the integration of multiple steps into enzyme catalyzed cascade processes [49].
- Co-immobilization of two or more enzymes allows multifunctional solid biocatalysts to catalyze such cascade processes [50].

Indeed, there is a paradigm shift in the manufacture of pharmaceuticals [51], agrochemicals [52], flavors and fragrances [53], and cosmetic ingredients [54] towards designing the biocatalytic processes as eco-benign, more cost effective than the more costly and less efficient classical syntheses, e.g., chemoenzymatic process for pregabalin synthesis [55], which is the active ingredient of the CNS drug Lyrica. The new pathway shows a substantial improvement in process efficiency by setting the stereocenter early, [56]) allowing the facile racemization and reuse of the wrong enantiomer in the synthesis. As against the traditional manufacturing process, the new process gives not only a higher yield but also five times reduction in the E factor from say 86 to 17.

4.7 GREEN MATRICES OF BIOTRANSFORMATIONS

Biocatalyzed processes can be carried out either with isolated enzymes or as whole cell bioconversions. Isolated enzymes are advantageous due to

Chemistry Matrices of Biotransformation

their purity in the cell. While, the use of whole cells is cheap, as it doesn't ask for the separation and purification of the enzyme. For dead cells, the E factors of the processes are essentially same; the waste in the form of cell debris is separated pre or post biotransformation. However, when growing microbial cells are used, say in fermentation processes, large amounts of biomass can be produced. The waste biomass so generated is normally easy to reuse, as animal feed or as a source of energy for the process. Several fermentation processes generate copious amounts of inorganic salts which poses a grave problem of waste disposal. The data pertaining to E factors for fermentation processes is not much available although it is badly required. From mass balances of some fermentation processes [57] can be used to calculate E factors, e.g., the E factor for the citric acid, a product of bulk fermentation is 1.4, which lies in the range <1–5, characteristic of bulk petrochemicals. Approximately 75% of the waste is contributed by calcium sulfate. Calcium hydroxide is used to regulate the pH, allowing calcium citrate, which is made to react with sulfuric acid to form citric acid and calcium sulfate. Inclusion of water in the calculation afforded an E factor of 17.

Another product of biomass conversion is bioethanol, derived from lignocellulose (wood) as the feedstock [58] and the E factor of ethanol derived from cellulose is 42. However, the major contributors to the waste are water about 36.8 kg/kg ethanol and carbon dioxide (4.1 kg/kg ethanol). The leftover is 1.1 kg/kg ethanol. Further, the generation of molasses (4 Kg), furfural (0.1 kg), and acetic acid (0.2 kg) as by-products are not considered in the calculation of E factor. It is observed that a cellulose-based ethanol plant processing 10,000 tons of lignocellulose feedstock/day to manufacture 870 tons of ethanol/day would produce wastewater (32 Mlit./day), that is, enough water to meet the requirements of a town having 300,000 residents. This water has many organic contaminants, the levels of which have to be reduced to the below ppm level so as to reuse the water which demands application of advanced and sophisticated industrial wastewater treatment methods necessarily. Small-volume fermentation processes for biopharmaceuticals may have very high E factors, as against the production of small-molecule drugs, and the use of water too is more.

The fermentative production of recombinant human insulin [57] has an E factor of 6600 and 50,000 without and with considering the amount of water used in the process respectively. The contributors to the waste and their amounts are tabulated in Table 4.2.

TABLE 4.2 The Contributors to the Waste and Their Amounts in Insulin Manufacture

Contributor	Amount in Kg/Kg of Insulin
Urea	1692
Acetic acid	1346
Formic acid	968
Phosphoric acid	713
Guanidine hydrochloride	445
Glucose	432
Sodium chloride	430
Acetonitrile	424
Sodium hydroxide	140

Biotransformations which use isolated enzymes have substantially high substrate concentrations, productivity with reduction in water usage against fermentations.

4.8 CASE STUDIES OF ENZYMATIC VIS-A-VIS CLASSICAL CHEMICAL PROCESSES

The following are the two case studies which involve:

- An enzyme-mediated process for the manufacture of an intermediate used for the synthesis of a cholesterol-reducing agent, Lipitor, the very first drug with worldwide sales turnover more than $10 billion; and
- An enzyme-mediated process for the production of a fatty acid ester used in personal care products.

4.8.1 ENZYME-MEDIATED SYNTHESIS OF AN INTERMEDIATE FOR ATORVASTATIN

Atorvastatin is an intermediate and an active ingredient of the cholesterol-reducing drug, Lipitor. As many as three-enzyme processes have been designed as a part of green-by-design, for the synthesis of a key intermediate in the manufacture of atorvastatin, the active ingredient of the cholesterol-lowering drug Lipitor [59, 60]. The process has been commercialized by Codexis. The first step involves the biocatalytic reduction of ethyl-4-chloroacetoacetate,

Chemistry Matrices of Biotransformation

using a ketoreductase (KRED) in presence of glucose and an NADP-dependent glucose dehydrogenase (GDH). The product, ethyl-4-chloro-3-hydroxybutyrate is obtainable with a 96% yield in isolation and >99.5% Enantiomeric excess (ee). It is followed by the application of halohydrin dehalogenase (HHDH) to catalyze a nucleophilic substitution of chloride with cyanide, using HCN under pH = 7 and room temperature (RT) conditions. Table 4.3 shows the directed evolution of KRED & GDH Biocatalyst.

All traditional synthetic routes to the hydroxy nitrile product (HN) require a standard SN2 substitution of halide with cyanide ion in alkaline medium at higher temperatures. As substrate and product both are sensitive to base reasonably, more amount by-product is likely to be formed. Further, the product exists as high-boiling oil, which demands a cumbersome high-vacuum fractional distillation to recover a qualitative product, leading to further losses in yield. The key in designing an economically and environmentally acceptable process for HN was to design a process for carrying out the cyanation reaction under mild conditions such as neutral pH, by using the enzyme, HHDH. Table 4.4 shows the evolution of the HHDH Biocatalyst. Coupling this with the chlorohydrin substrate synthesis using enantioselective KRED-catalyzed reduction of the respective keto ester, and cofactor regeneration with glucose/GDH, allowed an elegant two-step, three-enzyme process. However, the KRED and GDH are characterized by low activities and more enzyme loadings to maintain an economically viable reaction rate and are accompanied by cumbersome emulsion formation in downstream processing. Therefore, the analytical yield is >99% though, the practical yield is 85%.

To enable commercialization of the process on mass scale, the enzyme loadings should be reduced drastically and is achieved by *in-vitro* evolution (DNA shuffling technique) [61] to enhance the activity and stability of KRED and GDH while keeping intact the enantioselectivity exhibited by the KRED.

Many generations of DNA shuffling has led to the improvement in GDH activity by 13 folds and KRED activity by 7 folds keeping the enantio-selectivity at >99.5%. The use of modified enzymes has made it possible to complete the reaction in 8 h with the rise in substrate loading to 160 g/L and accompanied by reduction in enzyme loadings and no emulsion problems. Separation of phases require <1 min and offered the chlorohydrin >95% yield in isolation and with >99.9% ee. Further, the activity of the wild-type HHDH during the cyanation reaction is very poor and the enzyme exhibits drastic product inhibition and poor stability under process conditions. Due to large enzyme loadings, downstream processing becomes quite challenging. However, several iterative rounds of DNA shuffling, the inhibition gets reduced reasonably and in turn improvement in the HHDH activity by more than 2500-folds against wild-type enzyme. The greenness of the process is evaluated in accordance with the 12 principles of green chemistry [62].

TABLE 4.3 Directed Evolution of KRED & GDH Biocatalyst

Parameter	Process Design	Process	
		Initial	Final
Reactants loading (g/L)	159	78	160
Reaction time (h)	< 10	24	9
Enzyme loading (g/L)	< 1	9	0.9
Product yield (%)	> 90	84	96
Purity (%)	> 97	> 97	> 97
% Enantiomeric excess of product	> 99.3	> 99.3	> 99.6
Phase separation time (min)	< 11	> 63	< 1
Catalyst yield (G product/g catalyst)	> 159	9	177
Space-time yield (g product/L/	> 380	81	480

TABLE 4.4 Evolution of the HHDH Biocatalyst

Parameter	Process Design	Process	
		Initial	Final
Reactants loading (g/L)	> 120	24	141
Reaction time (h)	9	72	6
Enzyme loading (g/L)	1.5	30	1.2
Product yield (%)	> 90	69	92
Product purity (%)	> 96	> 96	> 96
%Enantiomeric excess of product	> 99.3	> 99.3	> 99.3
Phase separation time (min)	<10	> 63	<1
Catalyst yield(G product/g catalyst)	81	0.9	117
Space-time yield (g product/L/	> 360	9	672

4.9 APPLICATION OF GREEN CHEMISTRY PRINCIPLES

4.9.1 WASTE PREVENTION

The biocatalytic reactions which are highly selective reduce waste substantially. They bring about conversion of raw material into a product with >90%

Chemistry Matrices of Biotransformation 101

isolated yield with >98% chemical purity and an enantiomeric excess of >99.9%. The butyl acetate and ethyl acetate solvents, used in the extraction of the product, throughout the process are recycled with an efficiency of 85%. The E factor (kilogram waste per kilogram product) for the process at large is 5.8 and 18 if amount of water used is not used or used respectively in the calculation of the E factor. The main waste streams are aqueous and directly biodegradable.

The main contributors to the E factor are tabulated in Table 4.5.

TABLE 4.5 Contributors to E Factor

Waste	Quantity (Kg/Kg of HN)	% Contribution to E	
		Excluding Water	**Including Water**
Substrate losses (%)	0.09	<3%	<1%
Triethanolamine	0.03	<1%	<1%
NaCl and Na_2SO_4	1.29	24%	6%
Na gluconate	1.44	24%	9%
BuOAc (85% recycle)	0.45	9%	0.3%
EtOAc (85% recycle)	2.7	45%	15%
Enzymes	0.024	<1%	<1%
NADP	0.006	0.1%	< 0.1%
Water	12.24	—	69%
E factor	5.7	—	—

4.9.2 ATOM EFFICIENCY

The use of glucose as the reducing agent for cofactor regeneration is cheap but not atom economical (45%) in particular. However, glucose is a renewable resource while the gluconate coproduct is completely biodegradable.

4.9.3 LESS OR NONHAZARDOUS CHEMICAL SYNTHESIS

The reduction reaction uses starting materials which are nontoxic to both human and environmental health. It discourages the use of potentially hazardous hydrogen and heavy metal catalysts and subsequently obviates concern for their remediation from waste streams and product contamination. However, cyanide is used more efficiently in all practical routes to HN, and under milder conditions compared to earlier processes.

4.9.4 DESIGNING SAFER CHEMICALS

This principle is applicable to the HN product as the safe commercially available precursor for atorvastatin.

4.9.5 GREENER SOLVENTS AND AUXILIARIES

Safe and environmentally adoptable butyl acetate is used, with solvent water, in the biocatalytic reduction process and extraction of the HN product; no auxiliary's reagents are required.

4.9.6 SUSTAINABLE DESIGN PROVIDING FOR ENERGY EFFICIENCY AND CATALYSIS

As against earlier processes, which demand higher temperatures for the cyanation and high-pressure for hydrogenation, biocatalytic transformations are very efficient. The reactions are performed at or close to RT and pressure and pH 7 with no need of very high energy demands unlike high-vacuum distillation, leading to substantial energy conservation. The turnover numbers for the enzymes, KRED, and GDH >105 and >5×10⁴ for HHDH.

4.10 USE OF RENEWABLE FEEDSTOCKS AMENABLE TO BIODEGRADATION

The enzyme catalysts and the glucose co-substrate are obtained from renewable biodegradable raw materials. The by-products of the process are gluconate, NADP (the cofactor that moves from GDH to KRED), residual glucose, enzyme, and minerals, and the wastewater suit the needs of the bio-treatment. The process doesn't ask for derivatization and economy in steps, few unit operations against earlier processes, avoids the troublesome product distillation or the bisulfite-mediated separation of byproducts.

4.10.1 REAL-TIME ANALYSIS ALLOWING FOR POLLUTION PREVENTION AND INHERENTLY SAFER CHEMISTRY

The enzyme-mediated processes are carried out at neutral pH with computer-controlled addition of alkali. Gluconic acid HCl produced in the first and second steps is neutralized with aqueous NaOH and aqueous NaCN respectively, generating HCN (pKa~9) in situ. The pH and the total volume of the alkali added are recorded in real-time. Adding NaCN on demand reduces the overall concentration of HCN, allowing an inherently safer process.

In a nutshell, this process is an outstanding example of a biocatalytic process benign-by-design for the synthesis of a key pharmaceutical intermediate, the successful commercialization of which has been brought about by the application of modern protein engineering to optimize enzyme performance behavior. A Presidential Green Chemistry Challenge Award in 2006 was awarded to Codexis for designing of this process.

4.11 ENZYMATIC PRODUCTION OF AN EMOLLIENT ESTER

Biotechnological processes are applied in the cosmetics ingredients industry, wherein the "natural" label adds to the value of a product. The application of biocatalysis is advantageous even with comparatively fatty acid esters based on simple products like emollients [62], a conventional chemical catalyzed esterification compared with an enzyme-mediated one for the synthesis of the emollient ester, myristyl myristate used widely at industrial scale. The former involved the use of tin (II) oxalate used as a chemical catalyst under 240°C and 4 h as the conditions while the latter used an immobilized enzyme, Candida Antarctica lipase B, Novozyme 435, at 60°C for 12 h as shown in Figure 4.5.

The atom efficiency of the process is 96% and the E factor is <0.1 even when wastewater is included or otherwise. Consequently, the environmental impacts of the both processes were compared as a part of environmental life cycle analysis (LCA). The evaluation involved a comparison on the basis of the following five environmental impact parameters:

- Energy consumption;
- Global warming;
- Acidification;
- Nutrient enrichment; and
- Smog formation.

Substantial reductions in all categories were achieved. Table 4.6 shows biocatalytic esterification: Impact on Defined Environmental Parameters.

$$CH_3(CH_2)_{12}COOH + CH_3(CH_2)_{12}CH_2OH \xrightarrow[\text{Novozyme 435, 60°C}]{\text{Sn Oxalate 240°C}} CH_3(CH_2)_{12}CO_2(CH_2)_{13}CH_3 + H_2O$$

Atom efficiency 96

FIGURE 4.5 Chemo v/s biocatalytic production of myristyl myristate.

TABLE 4.6 Biocatalytic Esterification: Impact on Defined Environmental Parameters

Parameter	Units	Chemocatalytic	Biocatalytic	Savings
Energy	GJ	22.5	8.63	62%
Global warming	Kg CO_2 equivalent	1518	582	62%
Acidification	Kg SO_2 equivalent	10.58	1.31	88%
Eutrophication	Kg PO_4 equivalent	0.86	0.24	74%
Smog formation	Kg C_2H_4 equivalent	0.49	0.12	76%

Energy consumption was minimized by >60% and the generation of unwanted pollutants by ≈90%. Substitution of an environmentally unacceptable tin catalyst by an enzyme which can be used under considerably milder conditions are the key factors instrumental in making the biocatalytic process exceptionally more eco-benign. Moreover apart from other advantages, the higher product quality in the biocatalytic process, attributed to the much milder reaction conditions is quite beneficial which provides for circumvention of purification steps thereby resulting into simple downstream processing. Thus, better product quality and process simplification is the comparative advantages of the biocatalytic over the chemocatalytic process [62].

4.12 CONCLUSION

The road to green and sustainable chemicals manufacture hopefully is a catalytic one and employing meaningful green metrics for quantifying true reaction efficiency for monitoring progress of the processes. The green matrices such as AE and E factors are complementary and widely accepted metrics for the purpose of greening the syntheses. Combined with LCA, they offer a sound basis for assessing several processes and products in reference to their greenness and sustainability. Various sub-disciplines of catalysis—homogeneous, heterogeneous, organo-catalysis, and bio-catalysis will have a major role to play in the advent for sustainable technologies. Bio-catalysis, typically, has many advantages, namely mild reaction conditions using a catalyst that is biocompatible, biodegradable, and obtained from renewable materials. Processes are economical and highly selective amenable to higher product quality and reduced waste formation combined with superior economics. Added advantages also accrue from process intensification due to the integration of bio-catalytic steps with a cascading effect. Finally, catalysis in general, and bio-catalysis, in particular, will play a vital role in the shift from a global economy which is heavily reliant upon conventional fossil fuels derived feedstocks to a sustainable one based on renewable resources.

KEYWORDS

- **catalysis**
- **effective mass yield**
- **intensification**
- **life cycle assessment**
- **matrices**
- **reaction mass efficiency**

REFERENCES

1. Perkin, W. H., (1984). *J. Chem. Soc., 1862*, 232, Brit. Pat., 1856.
2. Peterson, D. H., & Murray, H. C., (1952). *J. Am. Chem. Soc., 71*, 1871–1872.
3. Holton, R. A., Biediger, R. J., & Boatman, P. D., (1995). In: Suffness, M., (ed.), *Taxol®:Science and Applications* (p. 97). Boca Raton (FL). CRC Press.
4. Sheldon, R. A., (2000). *CR Acad. Sci. Paris, IIc. Chim/Chem., 3*, 541–551.
5. Brundtland, C. G., (1987). *Our Common Future*. Oxford: The World Commission on Environmental Development, Oxford University Press.
6. Anastas, P., & Warner, J. C., (1998). *Green Chemistry: Theory and Practice*. Oxford University Press, Oxford.
7. Anastas, P. T., & Farris, C. A., (1994). *Benign by Design: Alternative Synthetic Design for Pollution Prevention, ACS Symposium Series No. 577*. Washington (DC): American Chemical Society.
8. Anastas, P. T., & Zimmerman, J. B., (2006). In: Abrahams, M. A., (ed.), *Sustainability Science and Engineering Defining Principles* (pp. 11–32). Elsevier, Amsterdam, Netherlands.
9. Tang, S. L. Y., Smith, R. L., & Poliakoff, M., (2005). *Green Chem., 7,* 761.
10. Sheldon, R. A., (1992). *Chem. Ind.,* (p. 903). London; (b) see also Sheldon, R. A., (1997). *Chem. Ind.,* (p. 12). London.

11. Sheldon, R. A., (1992). In: Sawyer, D. T., & Martell, A. E., (eds.), *Industrial Environmental Chemistry* (pp. 99–119). New York: Plenum.
12. Sheldon, R. A., (1993). In: Weijnen, M. P. C., & Drinkenburg, A. A. H., (eds.), *Precision Process Technology* (pp. 125–138). Dordrecht: Kluwer.
13. Sheldon, R. A., (1994). *Chemtech. March.,* 38–47.
14. (a) Sheldon, R. A., (1996). *J. Mol. Catal. A Chem., 107,* 75–83; (b) Sheldon, R. A., (1997). *J. Chem. Technol. Biotechnol., 68,* 381–388.
15. Sheldon, R. A., (2000). *Pure Appl. Chem., 72,* 1233–1246.
16. Trost, B. M., (1991). *Science, 254,* 1471–1477.
17. Trost, B. M., (1995). *Angew. Chem. Int. Ed., 34,* 259–281.
18. Caruana, C. M., (1991). *Chem. Eng. Prog., 87*(12), 11.
19. Sheldon, R. A., (1983). *Chemicals from Synthesis Gas* (p. 15). Dordrecht: Reidel.
20. Sheldon, R. A., (1985). *Bull. Soc. Chim. Belg., 94,* 651.
21. Notari, B., (1998). *Stud Surf. Sci. Catal., 37,* 413–425.
22. (a) Alfonsi, K., Collberg, J., Dunn, P. J., Fevig, T., Jennings, S., Johnson, T. A., Kleine, H. P., et al., (2008). *Green Chem., 10,* 31; (b) see also Dugger, R. W., Ragan, J. A., & Brown, R. D. H., (2005). *Org. Proc. Res. Dev., 9,* 253–258.
23. Ritter, S. K., (2008). *C & EN,* 59–68.
24. Thayer, A. N., (2007). *C & EN,* 11–19.
25. Calvo-Flores, F. G., (2009). *Chem. Sus. Chem., 2,* 905–919.
26. Auge, J., (2008). *Green Chem., 10,* 225–231.
27. Andraos, J., (2005). *Org. Proc. Res. Dev., 9,* 149–163.
28. Constable, D. J. C., Curzons, A. D., & Cunningham, V. L., (2002). *Green Chem., 4,* 521–527.
29. (a) Curzons, A. D., Constable, D. J. C., Mortimer, D. N., & Cunningham, V. L., (2001). *Green Chem., 3,* 1–6; (b) Constable, D. J. C., Curzons, A. D., Freitas, D. S. L. M., Green, G. R., Hannah, R. E., Hayler, J. D., Kitteringham, J., et al., (2001). *Green Chem., 3,* 7–9.
30. Hudlicky, T., Frey, D. A., Koroniak, L., Claeboe, C. D., & Brammer, L. E., (1999). *Green Chem., 1,* 57–59.
31. Eissen, M., & Metzger, J. O., (2002). *Chem. Eur. J., 8,* 3580–3585.
32. Jimenez-Gonzalez, C., Curzons, A. D., constable, D. J. C., & Cunningham, V. L., (2004). *Int. J. LCA, 9,* 114–121.
33. Moretz-Sohn, Monteiro, J. G., De Queiroz, F. A., & Luiz De, M. O. J., (2009). *Clean Technol. Environ. Policy, 11,* 209–214, 459–472.
34. Li, C. J., & Trost, B. M., (2008). *Proc. Natl. Acad. Sci. USA., 105,* 13197–13202.
35. Sheldon, R. A., & Van, B. H., (2001). *Fine Chemicals through Heterogeneous Catalysis.* Weinheim: Wiley-VCH, Chapters 3–7.
36. (a) Thomas, J. M., Hernandez-garrido, J. C., & Bell, R. G., (2009). *Top Catal., 52,* 1630–1639; (b) Kaneda, K., Mizugaki, T., (2009). *Energy Environ. Sci., 2,* 655–673.
37. Jimenez-Gonzales, C., Curzons, A. D., Constable, D. J. C., & Cunningham, V. L., (2004). *Int. J. Life Cycle Assess, 9,* 115–121.
38. Lindstrom, U. M., (2007). *Organic Reactions in Water.* Oxford: Blackwell.
39. (a) Leitner, W., (2002). *Acc. Chem. Res., 35,* 746; (b) Licence, P., Ke, J., Sokolova, M., Ross, S. K., & Poliakoff, M., (2003). *Green Chem., 5,* 99; (c) Cole-Hamilton, D. J., (2006). *Adv. Synth. Catal., 348,* 1341.
40. (a) Horvath, I. T., & Rabai, J., (1994). *Science, 266,* 72; (b) Horvath, I. T., (1998). *Acc. Chem. Res., 31,* 641.

Chemistry Matrices of Biotransformation

41. (a) Sheldon, R. A., (2001). *Chem. Commun.,* 2399; (b) Parvulescu, V. I., & Hardacre, C., (2007). *Chem. Rev., 107,* 2615–2665.
42. Wender, P. A., Handy, S. T., & Wright, D. L., (1997). *Chem. Ind. (London),* 765–769.
43. Bruggink, A., Schoevaart, R., & Kieboom, T., (2003). *Org. Proc. Res. Dev., 7,* 622.
44. Sheldon, R. A., (2005). *Green Chem., 7,* 267–278.
45. Leitner, W., Seddon, K. R., & Wasserscheid, P., (2003). Special issue on green solvents for catalysis. *Green Chem.,* 5, 99–284.
46. Noyori, R., (2005). *Chem. Commun.,* 1807–1811.
47. (a) Arnold, F. A., (2009*). Curr. Opin. Chem. Biol., 13,* 3–9; (b) Turner, N. J., (2009). *Nature Chem. Biol., 5,* 567–577; (c) Alexeeva, M., Carr, R., & Turner, N. J., (2003). *Org. Biomol. Chem., 1,* 4133–4137; (d) Turner, N. J., (2003). *Trends Biotechnol., 21,* 474–478; (e) Luetz, S., Giver, L., & Lalonde, J., (2008). *Biotechnol. Bioeng., 101,* 647–653; (f) Reetz, M. T., (2009). *J. Org. Chem., 74,* 5767–5778; (g) Reetz, M. T., (2006). *Adv. Catal., 49,* 1–69; (h) Bershtein, S., & Tawfik, D. S., (2008). *Curr. Opin. Chem. Biol., 101,* 647–653.
48. Sheldon, R. A., (2007). *Adv. Synth. Catal., 349,* 1289–1307.
49. Sheldon, R. A., (2008). In: Garcia-Junceda, E., (ed.), *Multi-Step Enzyme Catalysis: Biotransformations and Chemoenzymatic Synthesis* (pp. 109–135). Weinheim: Wiley-VCH.
50. (a) Mateo, C., Chmura, A., Rustler, S., Van, R. F., Stolz, A., & Sheldon, R. A., (2006). *Tetrahedron: Asymmetry., 17,* 320–323; (b) Van, P. S., Van, R. F., & Sheldon, R. A., (2009). *Adv. Synth. Catal., 351,* 397–404.
51. (a) Tao, J., & Xu, J. H., (2009). *Curr. Opin. Chem. Biol., 13,* 43–50; (b) Ran, N., Zhao, L., Chen, Z., & Tao, J., (2007). *Green Chem., 9,* 1–14; (c) Tao, J., Zhao, L., & Ran, N., (2007). *Org. Proc. Res. Dev., 11,* 259–267; (d) Yazbeck, D. R., Martinez, C. A., Hu, S., & Tao, J., (2004). *Tetrahedron: Asymmetry, 15,* 2757–2763; (e) Pollard, D. J., & Woodley, J. M., (2006). *Trends Biotechnol., 25,* 66–73; (f) Hailes, H. C., Dalby, P. A., & Woodley, J. M., (2007). *J. Chem. Technol. Biotechnol., 82,* 1063–1066. (g) Patel, R., Hanson, R., Goswami, A., Nanduri, V., Banerjee, A., Donovan, M. J., Goldberg, S., Johnston, R., et al. (2003). *J. Ind. Microbial. Biotechnol., 30,* 252–259.
52. Aleu, J., Bustillo, A. J., Hernandez-Galan, R., & Collado, I. G., (2006). *Curr. Org. Chem., 10,* 2037–2054.
53. (a) Franssen, M. C. R., Alessandrini, L., & Terraneo, G., (2005). *Pure Appl. Chem., 77,* 273–279; (b) Vandamme, E. J., & Soetaert, W., (2002). *J. Chem. Technol. Biotechnol., 77,* 1323–1332; (c) Serra, S., Fuganti, C., & Brenna, E., (2005). *Trends Biotechnol., 23,* 193–198.
54. (a) Veit, T., (2004). *Eng Life Sci., 4,* 508–511; (b) Heinrichs, V., & Thum, O., (2005). *Lipid Technol., 17,* 82–87.
55. Martinez, C. A., Hu, S., Dumond, Y., Tao, J., Kellcher, P., & Tully, L., (2008). *Org. Proc. Res. Dev., 12,* 392–398.
56. Sheldon, R. A., (1993). *Chirotechnology: The Industrial Synthesis of Optically Active Compounds.* New York: Marcel Dekker.
57. Petrides, B., (2003). In: Harrison, R. G., Todd, P. W., Rudge, S. R., & Petrides, D., (eds.), *Bioseparations Science and Engineering.* Oxford University Press, Oxford, UK, Chapter 11.
58. Rinaldi, R., & Schuth, F., (2009). *Energy Environ Sci., 2,* 610–626.
59. Fox, R. J., Davis, S. C., Mundorff, E. C., Newman, L. M., Gavrilovic, V., Ma, S. K., Chung, L. M., et al., (2007). *Nat. Biotechnol., 25,* 338.

60. Ma, S. K., Gruber, J., Davis, C., Newman, L., Gray, D., Wang, A., Grate, J., et al., (2010). *Green Chem., 12*, 81–86.
61. Stemmer, W. P., (1994). *Nature, 370*, 389–391.
62. Thum, O., & Oxenbøll, K. M., (2008). *SOFW J., 134*, 44–47.

CHAPTER 5

Sustainable Chemistry and Pharmacy

HITESH V. SHAHARE[1] and SHWETA S. GEDAM[2]

[1]SNJB's ShrimanSureshdada Jain College of Pharmacy, Chandwad, Maharashtra, India, E-mail: hiteshshahare1@rediffmail.com

[2]Sandip Institute of Pharmaceutical Sciences, Mahiravani, Nasik, Maharashtra, India

ABSTRACT

Green chemistry is today's need to minimize the environmental pollution by man-made materials and the processes which are used to produce them. By application of these green principles involved in green chemistry, provides an opportunity for sustainable future and world development. The challenge in this is to face and maintain the green chemistry practice with their rules in the process. The use of these principles is now days extended from academic laboratories to industries. Green chemists used the green technique and its process design for generation of biodegradable or recyclable products, to create sustainable raw materials which help to prevent the waste. The growth of green chemistry will give positive benefits in various areas of business profit, culture, and protection of our earth.

5.1 INTRODUCTION

"Green Chemistry" is the chemistry world which explains the acceptable chemical processes and products which are suitable for sustainable environment. It includes education, research, and commercial application of entire technologies for production of chemicals and products. It is accepted widely and acts as an essential development in practice of chemistry, and is very essential for sustainable development.

- **Definition:** Green Chemistry is a set of principles which decrease or remove the creation or use of toxic substances in the process, manufacture, and its application with their products [2]. It has been promoted worldwide in various organizations and research centers.

This is a new approach in view of sustainable environment to minimize the threats to health and the environment, the application of chemical substances, synthesis, and processing of it. This new approach is also known as:

- Environmentally benign chemistry;
- Clean chemistry;
- Atom economy (AE);
- Benign-by-design chemistry.

Green chemistry network channel work system includes:

- Training courses for teachers;
- Promotional events for the general public;
- Technology transfer events;
- New undergraduate course material including practicals;
- Website for schools.

Green chemistry involves the series of reductions. These reductions are responsible to achieve the objective of benefits to environment, social, and economical improvements. It is focused on chemical manufacturing processes [3].

- **Goals of Green Chemistry:** Green chemistry includes the use of various green technologies which aims to Figure 5.1.

5.2 HISTORY

The green chemistry term was the first time used by Anastas, in 1991 to implement sustainable growth in chemistry and chemical technology used by industry, academia, and government. To facilitate the contact between governmental agencies and industrial corporations with universities and research institutes, the Green Chemistry Institute was formed in 20 countries in view to design and implement the new green technologies. In the 1990s,

Sustainable Chemistry and Pharmacy

the first books and journals of green chemistry were introduced, and these were sponsored by the Royal society of chemistry [1].

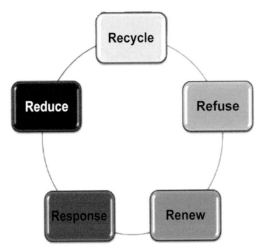

FIGURE 5.1 Goals of green chemistry.

5.3 PRINCIPLE

Paul Anastas and John C. Warner described the twelve principles of green chemistry. These principles focus on the basis of green technologies which help to attain the following basic requirements;

1. Design and utilization of processes which enhances the conversion of raw material to products, result in greater yield of product.
2. Enhance the use of substances which are non-toxic to environment and they should be obtained from environment including solvents.
3. Design of efficient processes related to use of energy.
4. Mostly avoid the formation of waste material in production process, so it will be helpful to reduce the problem of waste management. Somehow in the reaction, if it is not possible then it should be treated properly which will not harm to environment.

Green synthesis in pharmaceutical laboratories and industries is required for development of a sustainable environment (Figure 5.2) [4].

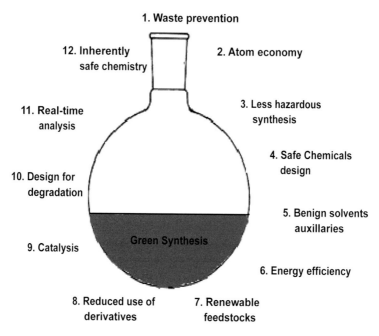

FIGURE 5.2 Principles of green chemistry.

On the platform of twelve principles of green chemistry, any chemical process can be modified to help for development of sustainable environment (Figure 5.3).

Sustainable chemistry involves the design, manufacture, and development and commercial application of non-hazardous chemicals and materials, application of manufacturing processes which helpful to eliminate or reduce the chemical risks for beneficial of human as well as environment.

This is more practicable by:

1. Maximizing use of more sustainable, renewable, and recycled substances;
2. Maximizing use of substitutes for rare materials;
3. Enhancing safe and more efficient manufacturing technology;
4. Minimizing environmental and health impacts from chemicals;
5. Optimizing the product and process design;
6. Enhancement to reuse and recycling of all materials and minimize the formation of waste (Table 5.1).

Sustainable Chemistry and Pharmacy

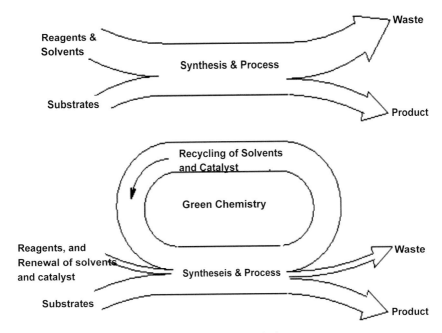

FIGURE 5.3 Conventional method verses green method.

TABLE 5.1 Twelve Principles of Green Chemistry

1.	Prevention	Prevent the waste formation during any process, so it will minimize the treatment or clean up of waste once it is formed
2.	Atom Economy	Synthetic methods should be designed well in which all raw materials converted into the final product
3.	Less Hazardous Chemical Synthesis	Synthetic methods should be designed by use of and produce the products which have minimum or no toxicity to health and environment
4.	Design Safe Chemicals	Design of chemical products which produce their maximum desired function and minimizing toxicity
5.	Safer Solvents and Auxiliaries	Use of auxiliary substances such as solvents, catalyst, separation agents, etc., should be not used unnecessarily and used only wherever it is essential
6.	Design for Energy Efficiency	The required energy for chemical processes should be recognized and it should not have any impact on the environment. The reactions for synthesis of materials should perform at ambient temperature and pressure

TABLE 5.1 *(Continued)*

7.	Use of Renewable Feedstock's	A raw material or feedstock use in reaction method should be of renewable form wherever technically practicable
8.	Reduce Derivatives	Derivatization by blocking groups, protection, and deprotection, temporary changes in physical and chemical method should avoid or it may be less. These steps require additional reagents and so generate more waste
9.	Catalysis	Catalytic reagents use for selectivity and so its use is superior to stoichiometric reagents
10.	Design for Degradation	The well design of chemical product may imparts the break-down of toxic degradation products at the end of reaction and it is not remain in the environment
11.	Real-Time Analysis for Pollution Prevention	Time to time development of analytical methodologies allows for monitoring of process and it helps to control the formation of toxic substances
12.	Inherently Safer Chemistry for Accident Prevention	The substances used in a chemical process used should be of less potential for chemical accidents and keep environment free from pollution

The German Federal Environmental Agency has prepared a document involving the principles of sustainable chemistry which is outlined as:

- **Qualitative Development:** The use of harmless substances is always better in view of humans and environment safety purpose. The production of long-life products will be always beneficial and will help in saving of resources.
- **Quantitative Development:** Natural resources utilization and its consumption rate if decrease and it should be renewable wherever possible. The minimizing, avoiding the entrapment of chemicals into the environment will serve to save the cost of process.
- **Comprehensive Life Cycle Assessment:** The analysis of raw material production, its processing, use, and disposal of chemicals with their products is essential to decrease the use of energy and various resources.
- **Action Instead of Reaction:** By avoiding the reactions of chemicals which are on the level of development and before to enter in market.
- **Economic Innovation:** The utilization of sustainable materials, their products and production methods produces a confidence in industrial users and private consumers to the customers from the public sector result in competitive advantages which lead to economic growth.

Sustainable Chemistry and Pharmacy

These all identified variables on sustainable chemistry which move the entire industry for safety production in view of improving environmental protection, consumer safety, health in occupational area and safety by removing the chemicals hazards (Table 5.2) [4, 5].

TABLE 5.2 Five Things to Make Your Laboratory More Sustainable

1.	Close the fume hoods when it is not in use, to reduce the use of energy
2.	Carry experiments at microscale to minimize the waste
3.	Switch to green solvents such as use of 2-methyl tetrahydrofuran which replace the methylene chloride and replacement of 1,4-dioxane, tetrahydrofuran, ether, by cylopentyl methyl ether
4.	Neutralization of reactions and chemicals like basic phosphate-buffered of HPLC waste to pH 7 and free to through in drainage
5.	Recyclization of materials such as electronics, ice packs, packaging materials, pipette tip boxes and cartridges

5.4 ENVIRONMENTALLY FRIENDLY TECHNOLOGIES

In outlook of objective of Green Chemistry, a pool of technologies are widely studied and used in seeking sustainable environment (Figure 5.4) [6].

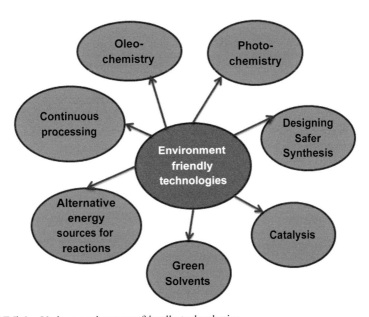

FIGURE 5.4 Various environment friendly technologies.

5.4.1 DESIGNING SAFER SYNTHESIS

The chemists design safer synthetic process in consideration to reduce the waste from solvents and added reagents. Another approach to reduce the hazards and toxicity of reagents is by running reactions on a small, continuous process.

5.4.2 CATALYSIS

The chemists want to build a fast, long-lived, and highly selective catalyst which will work in mild conditions and so will be helpful for a sustainable environment. Based on selectivity characteristics of a catalyst and also its regeneration property after a reaction, a catalyst can perform several transformations which allow scientists to earn high yields from a reaction in which a relatively small amount of catalyst has been used.

The transition metals like platinum, ruthenium, and palladium are used to build carbon-carbon bond-forming reactions.

In the research of nanoscience, provides insights into building green metal catalysts. The Gold-palladium nanocatalysts generate hydrogen peroxide, an oxidant for green chemistry, from hydrogen and oxygen. By attachment of catalyst to magnetic nanoparticles makes it an easy to separate and recycle the catalyst after a reaction process.

Acid catalysts attached to silica reduce the aqueous waste generated by quenching and neutralizing a reaction.

Phase transfer catalysts mostly ammonium salts contain an insoluble reactant between the organic and aqueous phases in a mixture.

Enzymes are commonly used as catalysts particularly in pharmaceuticals industry, because they works at intemperate temperature and pressure in water which makes them safe to handle on a process scale [7, 8].

5.4.3 GREEN SOLVENTS

Finding a new green solvent, in academic laboratories is targeted for green chemistry. The use of liquids like water, ionic liquids, and supercritical carbon dioxide (CO_2) plays a very important role:

- Water is not suitable solvent for performing a number of organic reactions because of many reagents are insoluble in water. But in some cases, such as aqueous Diels-Alder reaction between cyclopentadiene

Sustainable Chemistry and Pharmacy

and butanone, takes place very rapidly in water instead of organic solvents like methanol. Running a reaction in water makes it possible to use enzyme catalysts as well [9].

- Supercritical CO_2 shows the properties in between of a gas and liquid. It is s commonly used as a dry cleaning solvent. Scientists can control the compounds which dissolve in the supercritical fluid through its density adjustment to get its polarity. Because of volatility of CO_2, it is possible to remove easily after a reaction [10].
- Ionic liquids are used as green solvent. These liquid slats contain positive and negative ions controls the solubility and guide a reaction towards specificity. These liquids work as solvents for performing variety of reactions and they also designed to catalyze the reactions [11].

5.4.4   ALTERNATIVE ENERGY SOURCES FOR REACTIONS

When chemists perform a reaction by providing heat, simultaneously cool the solvent vapors coming out from reaction, the resulting droplets fall into the reaction so that the reaction never runs dry. This fact requires energy to both heat and to cool the reaction, in addition it also requires constant supply of water through a condenser. By outlook of this, chemist's finds new energy sources to drive reactions in view of green environment.

- Reactions designed by using microwave can help to maintain sustainable environment. Microwave-assisted reactions re possible in water at a small scale with accelerated rates due to the temperature and pressure effects [12].
- Reactions designed by using ultrasound sonication can help to maintain sustainable environment. It is another energy source in view of green chemistry with useful applications such as protecting hydroxyls on sugars, deprotecting an amine, or reducing α,β-unsaturated ketone in a steroid. The sound waves provide areas of high and low pressure (LP), as they travel through liquid. In the low-pressure areas, Bubbles are forming and collapse when they reach to high-pressure regions [13].

5.4.5   CONTINUOUS PROCESSING

A conventional synthesis in academic laboratories, research centers or in industries proceeds through a number of steps and reactors. It generates a large amount of solvent and chemical waste. The application of continuous processing method carries production in a steady flow of stream of liquid through addition of reagents. This application meets the nine amongst the twelve principles of green chemistry. These experiments are performed on the micro scale by pumping reactants through small channels. Continuous processes can also be designed to run short and multistep synthesis [14].

Example; A Grignard reaction is the first method in a pharmaceutical synthesis that needs stoichiometric control on Grignard reagent with ketone compound to avoid the production of byproducts and to prevent loss. The monitoring system measures and controls the flow-rate of the Grignard reagent into the reaction which helps to maintain the accurate stoichiometry.

5.4.6 OLEOCHEMISTRY (BIOLOGICAL SUBSTRATES)

Fats and oils obtained from plants and animals are used as oleochemical raw materials and exist as a novel source for chemicals. A series of raw materials from these sources are available in market. They are having number of applications in cosmetics, polymers, lubricating oils, and other products [15, 16].

5.4.7 PHOTOCHEMISTRY (CHEMICAL PROCESSES THROUGH LIGHT)

Light from source ultraviolet (UV) and visible are an important catalyst for performing many kinds of reactions. These photochemical methods are replacing toxic metals and reagents in many reactions [17].

5.5 THE BENEFITS OF GREEN CHEMISTRY

The utilization of principles of green chemistry for the development of a sustainable environment leads to beneficial for human health, environment, and the economy [18, 19].

5.5.1 HUMAN HEALTH

- **Clean Air:** Small release of toxic chemicals in air, result in less damage to lungs.

Sustainable Chemistry and Pharmacy

- **Cleaner Water:** Small release of toxic chemical wastes in water, result in pure drinking and recreational water.
- **Maximum Safety for Workers in Chemical Industry:** Small use of toxic chemicals will require less personal protective equipment and it will be less potential for accidents to occur such as fires or explosions.
- **Safer Consumer Products:** New safer products will be available to use, Pharmaceutical drugs will be prepared with minimum waste, and more pure form will be available.
- **Safer Food:** The removal of toxic chemicals which enter through the food; utilization of pesticides which are toxic to pests and itself degrade rapidly after use.

5.5.2 ENVIRONMENT

- Many chemicals after their function, they accumulate in environment by release such as pesticides, some by unintended releases including emissions during manufacturing, or some by disposal. The use of green chemicals either degrades to innocuous products or they are recovered for further use.
- It will prevent from smog formation and global warming.
- It will result in less chemical disruption of ecosystems.

5.5.3 ECONOMY AND BUSINESS

- Green chemistry principles will help to provide higher yields from chemical reactions.
- It will allow fewer synthetic steps, rapid production of products, maximizing the plant capacity and saving energy.
- Reduction in formation of waste during reaction, eliminating costly remediation, and remove hazardous waste disposal.
- It will allow the utilization of waste products instead of a purchased feedstock.
- Less products will be required to achieve the same function because of purity.
- The decreased use of petroleum products and avoiding their hazards will maximize the economy.
- Use of safer-product label to enhance consumer sales.

5.6 CONCLUSION

Green chemistry is extended from academic laboratories to the industry in the outlook of minimizing the costs, and also for environmental, health, and safety risks. Applications of principles involved in green technology are identified from small to large scales, by choosing ingredients for reactions, which minimizes the waste and risk to quantify the process efficiency.

The principles of green chemistry were focused on:

- Process design;
- During production, researcher to follow the energy use;
- Search for sustainable raw materials;
- To build biodegradable or recyclable products to prevent waste.

In view of developing sustainability, companies need to evaluate their processes and must incorporate green techniques principles in business monitoring. Regulations while drug production and the investment in existing chemical plants, minimize the development and applicability of up growing technologies [20].

KEYWORDS

- **carbon dioxide**
- **green future**
- **green principles**
- **green technology**
- **photochemistry**
- **sustainable chemistry**

REFERENCES

1. Anastas, P., & Eghbali N., (2010). Green Chemistry: *Principles and Practice. Chem. Soc. Rev., 39,* 301-312.
2. Anastas, P. T., & Hovarsth, I. T., (2007). Innovations and green chemistry. *Chem. Rev., 107,* 2169.
3. Sarbjeet, S. G., Sheelaa, M. A., Smriti, K., & Rajeev, K. S., (2012). A focus and review on the advancement of green chemistry. *Indo Global Journal of Pharmaceutical Sciences, 2*(4), 397–408.

Sustainable Chemistry and Pharmacy

4. Anita, I., Ana, D., Anita, M. B., & Stanislava, T., (2017). Review of 12 principles of green chemistry in practice. *International Journal of Sustainable and Green Energy*, *6*(3), 39–48.

5. Berkeley, W. C., & Zhang, J., (2009). Green process chemistry in the pharmaceutical industry. *Green Chemistry Letters and Reviews*, *2*(4), 193–211.

6. Jinliang, S. B. H., (2015). Green chemistry: A tool for the sustainable development of the chemical industry. *National Science Review*, *2*(3), 255–256.

7. Trost, B. M., (1995). Atom economy-A challenge for organic synthesis: Homogeneous catalysis leads the way. *Angew. Chem. Int. Ed.*, *34*, 259.

8. Carlo, L., & Angelo, V., (2011). Examples of heterogeneous catalytic processes for fine chemistry. *Green Chem.*, *13*, 1941.

9. Sheldon, R. A., (2005). Green solvents for sustainable organic synthesis: State of the art. *Green Chem.*, *7*, 267.

10. Sonali, R. S., (2015). Green chemistry. Green solvents and alternative techniques in org. synthesis. *International Journal of Chemical and Physical Science*, *1*, 516-520.

11. Schäffner, B., Verevkin, S. P., & Börner, A., (2012). Green solvents for synthesis and catalysis. Organic carbonates, *Chem. Unserer Zeit.*, *43*, 12–21.

12. Zovinka, E. P., & Stock, A. E., (2010). Microwave instruments: Green machines for green chemistry. *J. Chem. Educ.*, *87*, 350–352.

13. Ley, S. V., & Baxendale, I. R., (2002). New tools and concepts for modern organic synthesis. *Nat. Rev. Drug Discov.*, *1*(8), 573–586.

14. Vekariya, R. L., (2017). A review of ionic liquids: Applications toward catalytic organic transformations. *Journal of Molecular Liquids*, *227*, 44–60.

15. Li Ch, J., & Trost, B. M., (2008). Green chemistry for chemical synthesis. *Proc. Natl. Acad. Sci.*, *105*(36), 13197–13202.

16. Rajak, H., & Mishra, P., (2004). Microwave-assisted combinatorial chemistry: The potential approach for acceleration of drug discovery, *J. Sci. Ind. Res.*, *63*(8), 641–654.

17. Marek, T., Mariusz, M., Agnieszka, G., & Jacek, N., (2015). Green chemistry metrics with special reference to green analytical chemistry. *Molecules*, *20*, 10928–10946.

18. Alka, K., (2015). Review on green chemistry and its application. *International Journal of Research-Granthaalayah, 3*, 1–3.

19. Joshi, U. J., Gokhale, K. M., & Kanitkar, A. P., (2011). Green chemistry: Need for the hour. *Ind. J. Edu. Res.*, 168–174.

20. Bina, R., Raaz, M., & Chauhan, A. K., & Upma, S., (2012). Potentiality of green chemistry for future perspectives. *IJPCS, 1*(1), 1–6.

CHAPTER 6

Magnetism, Polyoxometalates, Layered Materials, and Graphene

FRANCISCO TORRENS[1] and GLORIA CASTELLANO[2]

[1]*Institute for Molecular Science, University of Valencia, PO Box 22085, E–46071 Valencia, Spain, E-mail: torrens@uv.es*

[2]*Department of Experimental Sciences and Mathematics, Faculty of Veterinary and Experimental Sciences, Valencia Catholic University Saint Vincent Martyr, Guillem de Castro-94, E–46001 Valencia, Spain*

ABSTRACT

It was pointed out the relation between the measurement that the two main molar magnetic susceptibilities (MSs) of graphite (-5×10^{-6} in the plane, -275×10^{-6} normal to the plane) and the theory of metallic conduction, which is shown by the graphite. Magnetic anisotropy (MA) and average magnetic susceptibility of graphite are reported to decay with decaying size of crystal grain. Graphite is included with metals because it shows metallic conductivity and the magnetic properties are, to a remarkable degree, similar to those found in some metals. The atomic magnetic susceptibility of graphite at room temperature (RT) is $\chi_A = 42\times10^{-6}$. Attention was drawn to the fact that graphite presents a large diamagnetic anisotropy. Transition metal dichalcogenides MX_2 (M = Ti, V, Zr, Nb, Ta; oxidation states (OSs) 5, 4, X = S, Se, Te) are studied for superconductors (SCs). The periodic table of the elements (PTE) of MX_2 for SCs is presented in this work. The nanoscale (NS) is only compatible with the external layer(s) of the bulk material. The history of electronics begins with Ge, then passes to Si and, finally, to C (with OSs 4, –4). The PTE of the history of electronics is presented in this study.

6.1 INTRODUCTION

Setting the scene: magnetism, diamagnetism (DM) and magnetic anisotropy (MA) of molecular crystals in graphite, metallic DM, and paramagnetism (PM) in graphite, polyoxometalate (POM)-based catalysts for carbon dioxide CO_2 conversion, physics of two-dimensional (2D) materials and graphene (GR) as the perfect carbon. It was pointed out the relation between the measurement that the two main molar magnetic susceptibilities (MSs) of graphite (as -5×10^{-6} in the plane and -275×10^{-6} normal to the plane) and the theory of metallic conduction, which is, of course, shown by graphite. It is interesting that MA and average MS of graphite are reported to decay with decaying size of crystal grain. Graphite is included with metals because it shows metallic conductivity and because the magnetic properties are, to a remarkable degree, similar to those found in some metals. The atomic MS of graphite (polycrystalline) at room temperature (RT) is $\chi_A = 42 \times 10^{-6}$. Attention was drawn to the fact that graphite presents a large DM MA (DMA).

Transition metal dichalcogenides (TMDs) MX_2 (M = Ti, V, Zr, Nb, and Ta; oxidation states (OSs) of 5 and 4, X = S, Se, and Te) are studied for superconductors (SCs). The periodic table of the elements (PTE) of MX_2 for SCs is presented in this work. The nanoscale (NS) is only compatible with the external layer(s) of the bulk material. The history of electronics began with Ge, then passed to Si and, finally, to C (with OSs of 4 and –4). The PTE of the history of electronics is presented in this study. Layered double hydroxides (LDHs) were reviewed [1].

In earlier publications, it was informed the effects of type, size, and elliptical deformation on molecular polarizabilities of model single-wall carbon nanotubes (SWNTs) from atomic increments [2–5], SWNTs periodic properties and table based on the chiral vector [6, 7], calculations on SWNTs solvents, co-solvents, *cyclo*pyranoses [8–11] and organic-solvent dispersions [12, 13], packing effect on cluster nature of SWNTs solvation features [14], information entropy analysis [15], cluster origin of SWNTs transfer phenomena [16], asymptotic analysis of coagulation-fragmentation equations of SWNT clusters [17], properties of fullerite and symmetric C-forms, similarity laws [18], fullerite crystal thermodynamic characteristics, law of corresponding states [19], cluster nature of nanohorns (SWNHs) solvent features [20], SWNTs (co-)solvent selection, *best* solvents, acids, superacids, host-guest inclusion complexes [21], C/BC_2N/BN fullerenes/SWNTs/ nanocones (SWNCs)/SWNHs/buds (SWNBs)/GRs cluster solvation models in organic solvents [22–28], elementary polarizability of Sc/*fullerene*/GR aggregates, di/GR-cation interactions [29], the world of materials from layered ones to conductive 2D metal-organic frameworks (MOFs) [30–34], porous materials (PMs), MOFs, and conduction [35]. The purpose of the present report is to review magnetism, DM, and MA of molecular crystals in graphite, metallic DM and PM in graphite, POM-based catalysts for CO_2 conversion, physics of 2D materials and GR as the perfect C. The aim of this work is to initiate a debate by suggesting a number of questions (Q), which can arise when addressing subjects of physics of 2D materials. It was provided, when possible, answers (A), and hypothesis (H) on these materials.

6.2 DIAMAGNETISM (DM) AND ANISOTROPY OF MOLECULAR CRYSTALS: GRAPHITE

Within the past few years, determination of main MSs in the crystal became a useful tool in structural chemistry, which is particularly true for aromatic compounds that usually show striking MA [36]. The information obtainable from magnetic data sometimes greatly simplifies determination of crystal structure by X-ray methods. The MA that is observed experimentally is because of, first, MA of the individual molecules and, second, their orientation in the crystal. If any two of the properties is known, or can be estimated, the third can be found. Lonsdale gives a semi-technical review of the field. Van Peype gives a theory of MA of cubic crystals at the absolute zero, and Neugebauer, a theory of DMA of acyclic molecules. However, the chief application is to aromatic compounds. Raman and Krishnan observed that the main MS of benzene normal to the plane of the ring is more than twice that in the plane. The phenomenon appears to be general in aromatic compounds and is sometimes developed to a pronounced degree. Classical theory of DM permits a reasonably accurate calculation of atomic radius from MS. Even for polyatomic molecules, a fair estimate of molecular radius may be made *via* Langeven's formula, which is definitely not true, however, for aromatic compounds. Failure of the theory to hold, even approximately, for aromatic compounds is related to the strong MA shown by the substances. Hückel and Ubbelohde stated qualitatively reasons for the behavior on the part of aromatic molecules, and Pauling, London, and Brooks, quantitatively. In aromatic compounds, the probability is that certain electrons are not restricted to individual atoms but flow all around the ring. According to Pauling, one (*p*) electron per aromatic C-atom is free to move between adjacent C-atoms under the influence of the impressed field. Molecules containing several condensed aromatic rings show a rise of main MS normal to the plane of the rings. It might be expected, therefore, that the effect would reach a maximum with graphite, and such is the case. Krishnan reports the two main molar MSs of graphite as -5×10^{-6} in the plane and -275×10^{-6} normal to the plane. Ubbelohde points out the relation between the effect and the theory of metallic conduction, which is, of course, shown by graphite. It is interesting that MA and average MS of graphite are reported to decay with decaying size of crystal grain. Squire and König report further theoretical work on DMA of aromatic molecules.

6.3 METALLIC DIA- AND PARAMAGNETISM (PM): GRAPHITE

Graphite is included in this section because it shows metallic conductivity and because the magnetic properties are, to a remarkable degree, similar to those found in some metals. The atomic MS of graphite (polycrystalline) at RT is $\chi_A = 42 \times 10^{-6}$. Attention was drawn to the fact that graphite presents a large DMA. The substance shows an outstanding example of free-electron DM that is restricted to one plane only. Ganguli and Krishnan determined MA over a wide temperature range for some exceptionally good crystals of Ceylon graphite. Perpendicular to the hexagonal axis $\chi = -0.5 \times 10^{-6}$ or nearly the same as for diamond. However, along the hexagonal axis, $\chi = -21.5 \times 10^{-6}$ at RT. Variation of MA with temperature is shown (*cf.* Figure 6.1), where $\chi_{\parallel} - \chi_{\perp} = \chi_e$, i.e., the difference in DM along the hexagonal axis and perpendicular to it, is plotted *vs.* temperature. It will be seen that at high temperatures (HTs, 1270 K), χ_e tends asymptotically to reach $-0.010/T$, while at low temperatures (LTs, 90 K), it approaches a temperature-independent value of -30×10^{-6}. It may be shown that the large DM along the hexagonal axis is because of one free electron per C-atom, and the electron is completely free in the basal plane but highly restricted in the perpendicular direction. Every layer of C-atoms may be thought of as a giant aromatic molecule, which, incidentally, constitutes proof that the bonds in aromatic molecules are not localized, i.e., it establishes the basic concept of the resonance theory as applied to such substances. In order to explain the non-appearance of free-electron PM in graphite, Ganguli, and Krishnan suggest that all occupied energy levels in graphite must contain pairs of electrons, and no such levels contain single electrons. It follows that the energy of spin coupling must be large compared to kT even at 1270 K. The abnormal DMA of graphite is largely destroyed by partial oxidation in nitric HNO_3 plus sulfuric H_2SO_4 acids to form *blue graphite*.

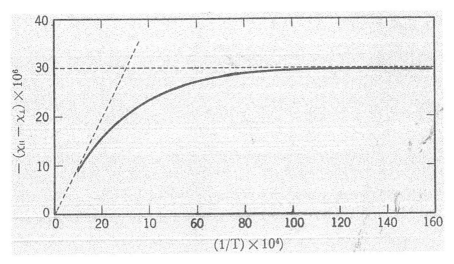

FIGURE 6.1 DMA $\chi_\parallel - \chi_\perp$ vs. temperature for a single crystal of graphite.
Source: Ref. [36].

6.4 POLYOXOMETALATE (POM)-BASED CATALYSTS FOR CO_2 CONVERSION

The POMs are a diverse class of anionic metal-oxo clusters with intriguing chemical and physical properties. Because of unrivaled versatility and structural variation, POMs were extensively utilized for catalysis for a plethora of reactions. Applications of POMs were reviewed as promising catalysts or co-catalysts for CO_2 conversion [e.g., CO_2 photo/electro reduction, as a carbonyl (C=O) source for the carbonylation process] [37]. A perspective on the potentiality in the field was proposed.

6.5 PHYSICS OF 2D MATERIALS WORKSHOP

Navarro-Moratalla organized *Physics of 2D Materials* Workshop [38]. Jarillo-Herrero raised Q on correlations/SC in magic-angle (MA) framework GR [39].

Q1. *High-temperature* (HT) phase diagram MA-twisted bilayer GR (TBG) [40]?

Lu raised the following questions on SCs, orbital magnets and quantum oscillations [41].

Q2. What should we expect if all these in homogeneities are lifted?
Q3. SC in-between integer fillings?
Q4. What did we learn about the origin of SC?
Q5. Do correlations give rise to SC?
Q6. Do correlations compete with SC?
Q7. Are there correlated states with lifted spin/valley degeneracy?
Q8. Magnetic phase transition at $v = -1$?
Q9. Implications of improved twist-angle homogeneity?

Valenzuela raised Q on spin-orbit effects at hot-carrier transport in GR-based devices [42].

Q10. Is the spin relaxation in GR fully understood?
Q11. Reference GR device: Is there the same probability for the electrons to go spin up or down?

Mañas Valero proposed questions/answer on old materials, new physics, and TaS_2 case [43].

Q12. Who does it know what it is?
Q13. SC MX_2?
A13. The TMDs MX_2 [M = Ti, V, Zr, Nb, Ta; OSs 5 and 4, X = S, Se, Te, (*cf.* Figure 6.2) for SCs.

He raised additional questions:

Q14. Why critical temperature (CT) is going down when layering?
Q15. TaS_2 (CT going up *vs.* layering) *vs.* MoS_2 (CT going down *vs.* layering)?
Q16. Charge density waves?
Q17. Other possible explanations?
Q18. Some other experiments?
Q19. Is TaS_2 magnetic?
Q20. The 2D molecular magnetism: Molecules?

Magnetism, Polyoxometalates, Layered Materials, and Graphene   129

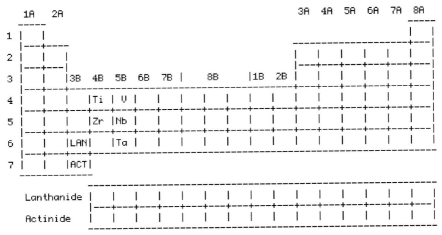

FIGURE 6.2 Periodic table of the elements of MX_2 for superconductors.

Molina-Sánchez raised Q on light-matter interaction in 2D ferromagnetic (FM) materials [44].

Q21. Absorption of circularly polarized light excitonic effects on CrI_3?

Fernandez Rossier proposed Q/A/H on CrI_3 monolayers/proximity effects/interlayer coupling [45].

- Q22. Why do we care?
- Q23. Why monolayers are FM?
- A23. Magnetic anisotropy.

- H1. (Mermin-Wagner). Theorem: continuous symmetries cannot be *spontaneously broken* at finite temperature in systems with sufficiently short-range interactions in dimensions $d \leq 2$.
- Q24. Why is there FM order?
- A24. Mermin-Wagner theorem.
- Q25. What is the origin of *magnetic anisotropy* effect (MAE) in CrI_3?
- Q26. Why is tunnel magnetoresistance so large?
- A26. GR Dirac cone resonant with, e.g., *spin orientation*.

Asensio proposed Q/A on materials chemical/electronic imaging for advanced nanotechnologies in energy/electronics [46].

Q27. Beyond GR?
A27. Other low-dimensional materials.
Q28. Are two 2D materials better than one?
Q29. GR Dirac band: Decoupled twisted layers?
Q30. ABA, AAA, ABC stacking?

She proposed the following conclusion (C):

C1. The NS is only compatible with the external layer(s) of the bulk material.

6.6 GRAPHENE (GR): THE PERFECT CARBON

Da Silva raised the following questions (Qs) on GR as the perfect C (*cf.* Figure 6.3) [47].

Q1. However... What can be manufactured with GR?
Q2. Which are such characteristics that make it so special?
Q3. Why was it announced that it would revolutionize the laboratories?
Q4. Do people already consume products with GR?
Q5. Is it true that it allows the development of intelligent clothing?

FIGURE 6.3 Graphene: The perfect carbon.

She presented electronics history beginning with Ge, then passing to Si and, finally, C (OSs 4, –4, cf. Figure 6.4).

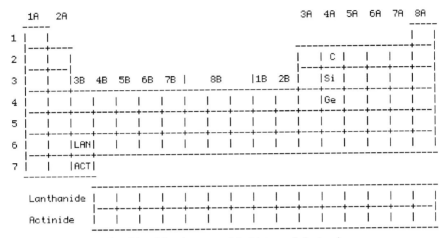

FIGURE 6.4 Periodic table of the elements of the history of electronics.

6.7 FINAL REMARKS

From the present results and discussion, the following final remarks can be drawn:

1. It was pointed out the relation between the measurement that the two main molar MSs of graphite (as -5×10^{-6} in the plane and -275×10^{-6} normal to the plane) and the theory of metallic conduction, which is, of course, shown by graphite. It is interesting that MA and average magnetic susceptibility of graphite are reported to decay with decaying size of crystal grain.
2. Graphite is included with metals because it shows metallic conductivity and because the magnetic properties are, to a remarkable degree, similar to those found in some metals. The atomic magnetic susceptibility of graphite (polycrystalline) at RT is $\chi_A = 42\times10^{-6}$. Attention was drawn to the fact that graphite presents a large diamagnetic anisotropy.
3. Transition metal dichalcogenides MX_2 (M = Ti, V, Zr, Nb, and Ta; OSs of 5 and 4, X = S, Se, and Te) are studied for SCs. The PTE of MX_2 for SCs is presented in this work.

4. The NS is only compatible with the external layer(s) of the bulk material.
5. The history of electronics began with Ge, then passed to Si and, finally, to C (with OSs of 4 and –4). The PTE of the history of electronics is presented in this study.

ACKNOWLEDGMENTS

The authors thank support from Fundacion Universidad Catolica de Valencia San Vicente Martir (Project No. 2019-217-001UCV).

KEYWORDS

- **carbon dioxide**
- **carbonyl source**
- **catalysis**
- **magnetochemistry**
- **periodic table of the elements**
- **photo/electro reduction**
- **two-dimensional material physics**

REFERENCES

1. Duan, X., & Evans, D. G., (2005). *Layered Double Hydroxides.* Springer: Berlin, Germany.
2. Torrens, F., (2003). Effect of elliptical deformation on molecular polarizabilities of model carbon nanotubes from atomic increments. *J. Nanosci. Nanotech., 3*, 313–318.
3. Torrens, F., (2004). Effect of size and deformation on polarizabilities of carbon nanotubes from atomic increments. *Future Generation Comput. Syst., 20*, 763–772.
4. Torrens, F., (2004). Effect of type, size, and deformation on polarizability of carbon nanotubes from atomic increments. *Nanotechnology, 15*, S259–S264.
5. Torrens, F., (2006). Corrigendum: Effect of type, size, and deformation on polarizability of carbon nanotubes from atomic increments. *Nanotechnology, 17*, 1541–1541.
6. Torrens, F., (2004). Periodic table of carbon nanotubes based on the chiral vector. *Internet Electron. J. Mol. Des., 3*, 514–527.
7. Torrens, F., (2005). Periodic properties of carbon nanotubes based on the chiral vector. *Internet Electron. J. Mol. Des., 4*, 59–81.

8. Torrens, F., (2005). Calculations on *cyclo*-pyranoses as co-solvents of single-wall carbon nanotubes. *Mol. Simul., 31*, 107–114.

9. Torrens, F., (2005). Calculations on solvents and co-solvents of single-wall carbon nanotubes: *Cyclo*-pyranoses. *J. Mol. Struct. (Theochem.), 757*, 183–191.

10. Torrens, F., (2005). Calculations on solvents and co-solvents of single-wall carbon nanotubes: *Cyclo*pyranoses. *Nanotechnology, 16*, S181–S189.

11. Torrens, F., (2005). Some calculations on single-wall carbon nanotubes. *Probl. Nonlin. Anal. Eng. Syst., 11*(2), 1–16.

12. Torrens, F., (2006). Calculations of organic-solvent dispersions of single-wall carbon nanotubes. *Int. J. Quantum Chem., 106*, 712–718.

13. Torrens, F., & Castellano, G., (2005). Cluster origin of the solubility of single-wall carbon nanotubes. *Comput. Lett., 1*, 331–336.

14. Torrens, F., & Castellano, G., (2007). Cluster nature of the solvation features of single-wall carbon nanotubes. *Curr. Res. Nanotech., 1*, 1–29.

15. Torrens, F., & Castellano, G., (2007). Effect of packing on the cluster nature of C nanotubes: An information entropy analysis. *Microelectron. J., 38*, 1109–1122.

16. Torrens, F., & Castellano, G., (2007). Cluster origin of the transfer phenomena of single-wall carbon nanotubes. *J. Comput. Theor. Nanosci., 4*, 588–603.

17. Torrens, F., & Castellano, G., (2007). Asymptotic analysis of coagulation-fragmentation equations of carbon nanotube clusters. *Nanoscale Res. Lett., 2*, 337–349.

18. Torrens, F., & Castellano, G., (2008). Properties of fullerite and other symmetric carbon forms: Similarity laws. *Symmetry Cult. Sci., 19*, 341–370.

19. Torrens, F., & Castellano, G., (2010). Fullerite crystal thermodynamic characteristics and the law of corresponding states. *J. Nanosci. Nanotechn., 10*, 1208–1222.

20. Torrens, F., & Castellano, G., (2010). Cluster nature of the solvent features of single-wall carbon nanohorns. *Int. J. Quantum Chem., 110*, 563–570.

21. Torrens, F., & Castellano, G., (2011). (Co-)solvent selection for single-wall carbon nanotubes: Best solvents, acids, superacids, and guest-host inclusion complexes. *Nanoscale, 3*, 2494–2510.

22. Torrens, F., & Castellano, G., (2012). *Bundlet* model for single-wall carbon nanotubes, nanocones and nanohorns. *Int. J. Chemoinf. Chem. Eng., 2*(1), 48–98.

23. Torrens, F., & Castellano, G., (2013). Solvent features of cluster single-wall C, BC$_2$N and BN nanotubes, cones, and horns. *Microelectron. Eng., 108*, 127–133.

24. Torrens, F., & Castellano, G., (2013). Corrigendum to: Solvent features of cluster single-wall C, BC$_2$N and BN nanotubes, cones, and horns. *Microelectron. Eng., 112*, 168–168.

25. Torrens, F., & Castellano, G., (2013). *Bundlet* model of single-wall carbon, BC$_2$N and BN nanotubes, cones, and horns in organic solvents. *J. Nanomater. Mol. Nanotech., 2*, 1000107–1–9.

26. Torrens, F., & Castellano, G., (2013). C-nanostructures cluster models in organic solvents: Fullerenes, tubes, buds and graphenes. *J. Chem. Chem. Eng., 7*, 1026–1035.

27. Torrens, F., & Castellano, G., (2014). Cluster solvation models of carbon nanostructures: Extension to fullerenes tubes and buds. *J. Mol. Model., 20*, 2263–1–9.

28. Torrens, F., & Castellano, G., (2014). Cluster model expanded to C-nanostructures: Fullerenes, tubes, graphenes and their buds. *Austin J. Nanomed. Nanotech., 2*(2), 7–1–7.

29. Torrens, F., & Castellano, G., (2013). Elementary polarizability of Sc/*fullerene/graphene* aggregates and di/graphene-cation interactions. *J. Nanomater. Mol. Nanotech., S1*, 001–1–8.

30. Torrens, F., & Castellano, G., (2018). Conductive layered metal-organic frameworks: A chemistry problem. *Int. J. Phys. Study Res., 1*(2), 42–42.
31. Torrens, F., & Castellano, G., (2019). Conductive two-dimensional nanomaterials: Metal-organic frameworks. *J. Appl. Phys. Nanotech., 2*(2), 52–52.
32. Torrens, Z. F., (2019). From layered materials to bidimensional metal-organic frameworks. *Nereis, 2019*(11), 65–80.
33. Torrens, F., & Castellano, G. (2018). Conductive layered metal–organic frameworks: A chemistry problem. *Int. J. Phys. Study Res., 1(2)*, 42–42.
34. Torrens, F., & Castellano, G. World of conductive two-dimensional metal-organic frameworks. In: Yaser, A., Z., Khullar, P., & Haghi, A. K., (eds.), *Green Materials and Environmental Chemistry: New Production Technologies, Unique Properties, and Applications.* Apple Academic-CRC: Waretown, NJ, in press.
35. Torrens, F., & Castellano, G. Porous materials, metal-organic frameworks and conduction. In: Esteso, M. A., Ribeiro, A. C. F., & Haghi, A. K., (eds.), *Chemistry and Chemical Engineering for Sustainable Development: Best Practices and Research Directions.* Apple Academic-CRC: Waretown, NJ, in press.
36. Selwood, P. W., (1943). *Magnetochemistry.* Interscience: New York, NY.
37. Cao, Y., Chen, Q., Shen, C., & He, L., (2019). Polyoxometalate-based catalysts for CO_2 conversion. *Molecules, 24*, 2069–1–26.
38. Navarro-Moratalla, E., (2019). *Book of Abstracts.* Physics of 2D Materials Workshop, València, Spain. Universitat de València: València, Spain.
39. Jarillo-Herrero, P., (2019). *Book of Abstracts* (pp. 1–1). Physics of 2D materials workshop, València, Spain, Universitat de València: València, Spain.
40. Cao, Y., Fatemi, V., Fang, S., Watanabe, K., Taniguchi, T., Kaxiras, E., & Jarillo-Herrero, P., (2018). Unconventional superconductivity in magic-angle graphene superlattices. *Nature (London), 556*, 43–50.
41. Lu, X., (2019). *Book of Abstracts.* Physics of 2D materials workshop, València, Spain. Universitat de València: València, Spain, O-3.
42. Valenzuela, S. O., (2019). *Book of Abstracts* (pp. 1–4). Physics of 2D materials workshop, València, Spain. Universitat de València: València, Spain.
43. Mañas, V. S., (2019). *Book of Abstracts* (pp. 1–5). Physics of 2D materials workshop, València, Spain, Universitat de València: València, Spain.
44. Molina-Sánchez, A., (2019). *Book of Abstracts* (pp. 1–7). Physics of 2D materials workshop, València, Spain. Universitat de València: València, Spain.
45. Fernandez, R. J., (2019). *Book of Abstracts* (pp. 1–9). Physics of 2D materials workshop, València, Spain. Universitat de València: València, Spain.
46. Asensio, M. C., (2019). *Book of Abstracts* (pp. 1–10). Physics of 2D materials workshop, València, Spain. Universitat de València: València, Spain.
47. Da Silva, E., (2019). *Personal Communication.*

CHAPTER 7

Plasma, Photo/Radiochemical Reactions, and Relativity Theories

FRANCISCO TORRENS[1] and GLORIA CASTELLANO[2]

[1]*Institute for Molecular Science, University of Valencia, PO Box 22085, E–46071 Valencia, Spain, E-mail: torrens@uv.es.*

[2]*Department of Experimental Sciences and Mathematics, Faculty of Veterinary and Experimental Sciences, Valencia Catholic University Saint Vincent Martyr, Guillem de Castro-94, E–46001 Valencia, Spain*

ABSTRACT

The quantum yield is the main characteristic of classifying photochemical reactions, because its value varies within broad limits. A feature of photochemical reactions is that their rate does not practically depend on the temperature. Photochemical and radiochemical reactions are responsible for processes going on with a rise in Gibbs function of the system, and decay in the entropy of the universe, i.e., non-spontaneous processes, which aspect points to the alluring prospect of using the reactions for a number of syntheses. The vacuum is not the nothingness; it is a substance $\varphi \neq 0$. If the vacuum is a substance, one can shake it; its vibrations are elementary particles: the Higgs. The standard models (SMs) of micro and macrocosmos burst. Einstein invented the general theory of relativity (GTR) and quantum mechanics (QM) wave-particle photon, but nobody knows how to unify them. In Einstein's blackboard, there is a lot to fill.

7.1 INTRODUCTION

Setting the scene: plasma, properties, ways of generation, applications, exercises, photochemical, and radiochemical reactions, exercise, solution, relativity theories, particle physics, accelerators, *zoos*, models, the perfect theory, Einstein, the vacuum and the nothingness.

In earlier publications, it was informed nuclear fusion, American nuclear cover-up in Spain after Palomares (Almería, Spain) disaster (1966) [1], *Manhattan Project*, *Atoms for Peace*, nuclear weapons, accidents [2], nuclear science, technology [3], history, concept, method, didactics of atomic and nuclear physics [4], gravitational waves, messengers of the universe and cosmology [5]. The aim of the present work is to review plasma, properties, ways of generation, applications, exercises, photochemical, and radiochemical reactions, solution, relativity theories, particle physics, accelerators, *zoos*, models, the perfect theory, Einstein, the vacuum and the nothingness. The goal of this work is to initiate a debate by suggesting a number of questions (Q) on relativity theories, particle physics, accelerators, *zoos*, models, the perfect theory, Einstein, the vacuum and the nothingness, and providing, when possible, answers (A) and hypotheses (H).

7.2 PLASMA

7.2.1 GENERAL

The solid (S), liquid (L), and gaseous (G) states of a substance are commonly known and habituary under normal conditions [6]. However, another state of aggregation is possible, i.e., a *plasma* (P). It was established that rise in the temperature leads to a rise in not only the kinetic energy of molecules (Maxwell's law), but also the energy of the particles forming a molecule. At sufficiently high temperatures (HTs), the bond between the particles inside a molecule weakens and breaks. The molecules decompose into ions, and the body acquires a new state: it becomes plasma, which are an ionized gas consisting of neutral molecules and atoms, cations, and anions, free electrons (e^-) and other particles. The assembly of every kind of the particles is represented as an individual component of the gas mixture forming the plasma [e.g., an assembly of neutrons (n^0) is considered as an n^0-gas, an assembly of protons (p^+), as a p^+-gas, an assembly of e^-, as an e^--gas]. The total number of cations and anions in plasma is such that, as a whole, it is electrically neutral.

7.2.2 PROPERTIES

The basic properties of a plasma are its HT (10^{3-9} K) and electric conductance because of the intensive motion of the free ions. No strict transition boundary

Plasma, Photo/Radiochemical Reactions, and Relativity Theories

(interface) exists between plasma and gas as it exists between a gas, liquid, and solid (crystalline) body. An HT plasma presents the properties of an ideal gas (IG) and obeys IG equation of state, which is why a plasma is called a *conducting gas*. In plasma physics, the temperature is expressed in electron-volts (1e V = 11,600 K), where the temperature characterizes the heat (kinetic) energy of the particles, which in a plasma, unlike gases, is not always identical for all its particles. If the mean energies of plasma particles regardless of their nature are identical, one has to do with *isothermal* plasma. The temperature of such plasma is measured in scores and hundreds of electron-volts, i.e., millions of Kelvin. An isothermal plasma is called *hot*, which motion of its particles confined in a strong magnetic field (MF) is constricted by MF force lines (MFLs). Such plasma is said to be magnetized. Its particles move vertically along MFLs, and the chaotic motion of the particles becomes ordered, which plasma can be restrained by an MF *wall*, pushed by an MF *piston* (MP), and retained in an MF *trap* (MT). If hot plasma is placed in a strong electric field (EF), i.e., a heavy current is passed *via* it, it will be compressed and stretch out into a *plasma column* or pinch. The methods of localizing hot plasma are based on the properties. No heat-resistant material can retain plasma: In a few fractions of a second, it burns *via* and melts any insulating material. Most usually when designing devices capable of retaining hot plasma, a toroidal chamber (doughnut) with MF coils is employed (*cf.* Figure 7.1), i.e., a giant hollow doughnut with the winding of electromagnets on it. The plasma in the hollow of the doughnut is picked up by MFLs and begins to travel in a circle along the chamber. It is compressed into a pinch at the center of the doughnut hollow, being separated from the walls. The vacuum formed between the plasma and the doughnut walls thermally insulates the plasma. Hot plasma can be retained in other ways *via* its electromagnetic (EM) properties. However, unresolved problems exist in the field, one of which is the instability of the plasma. Magnetized plasma can produce spikes directed across MFLs similar to the protuberances of Sun. Such a disturbance of hot plasma can destroy all the apparatus, retaining it in a single instant. The problem is not solved and experiments are mainly being conducted with a relatively stable rarefied plasma, whose density is 10^{20-21} particles·m^{-3}. Under a low pressure (LP), the exchange of energy between the particles results hampered. The heavier slow plasma particles readily give up their energy to the surrounding medium and present a lower energy than the lighter, fast particles that do not virtually exchange their energy with the surroundings. The temperature of different components of the plasma varies and the plasma is *non-isothermal*, which temperature of its e⁻ and *photon* gases is much higher than that of its ionic, molecular, radical, atomic, etc., gases, e.g., the temperature of e⁻-gas in Ne daylight lamps reaches 25,000 K, whereas the temperature because of the neutral particles and ions is close to the ambient temperature. The temperature of such plasma does not usually exceed 1,000 K, which is why it is called *cold* plasma.

FIGURE 7.1 Operating principle of a doughnut with magnetic coils.
Source: Ref. [6].

7.2.3 WAYS OF GENERATION

In order to obtain an H-plasma artificially, it is sufficient to heat the gas to 10^6 K, which ways of heating are diverse and depend on the kind of plasma that is to be obtained. In order to produce cold plasma, an electric discharge in a rarefied gas is sufficient: cold plasma is called a *gas-discharge* one. Such plasma appears (e.g., in Ne lamps, corona discharge, Ar plasma burners). Hot plasma can be produced only in MTs. The gas can be heated to the required temperature in the case by a heavy electric current, high-frequency (HF) discharges, or rapidly compressing the gas with an MP. Plasma obtained in the way is called *thermal*. The degree of ionization of the atoms of a substance in a thermal plasma approaches 100%. Completely ionized plasma exists (with thermal insulation) an infinitely long time. Gas-discharge plasma is ionized relatively poorly. It exists only with the constant action of an EF needed to replenish the stock of ions in the plasma, because the ions are continuously depleted as a result of collisions with neutral particles and the walls of the vessel.

7.2.4 APPLICATIONS

A hybrid plasma with a temperature of 400–11,000 K found the greatest technical application at present. It is employed in the hot working of metals (plasma, electric welding, etc.), in the refractory industry and metallurgy

(HF induction-plasma flame torches). Attention must be given to the use of plasma in chemistry (*plasmochemistry*). The plasma produced in a number of media (H_2, N_2, O_2, inert gases, etc.), allows endothermal reactions to be realized, such that under conventional conditions proceed slowly or even cannot proceed at all for thermodynamic reasons, e.g., NO_x is synthesized in O-plasma when producing HNO_3, metals are reduced from their ores in H-plasma, acetylene, and technical H_2 are produced from natural gas (NG) CH_4 in the plasma of an electric arc, unsaturated hydrocarbons (UHCs) are produced from petrol under similar conditions, etc. The high rate of reactions proceeding in a plasma made it possible to create highly productive small reactors (*plasmotrons*), e.g., a CH_4 plasmotron with an output of 25,000 t of acetylene a year presents a length of 65 cm and a diameter of 15 cm, which makes the production process much economical. In recent years, the use of cold plasma at a temperature below 400 K grew, which is because of the ability of many organic compounds to enter into chemical reactions with atomic gases, and active particles similar to free radicals (FRs) that form in large numbers in cold plasma. It is presumed that cold plasma will be able to replace catalysts. Great hopes are attached to hot plasma, whose use would facilitate the solution of one of the fundamental tasks of our time, i.e., the production of energy. A promising possibility of employing a plasma for the purpose is the creation of a magnetohydrodynamic (MHD) generator, which principle is based on the fact that when a stream of a plasma is passed at a high velocity *via* an MF with a high induction, an electric current is induced in the plasma that could be directed to an external load, which will create a heat power plant without any turbines. Of major interest for power engineering is controlled thermonuclear reactions, i.e., processes of the fusion of light atomic nuclei into heavy ones, which processes occur at an HT ($\sim 10^6$K) and are attended by the evolution of a tremendous amount of energy. Such reactions are constantly proceeding in stars, e.g., Sun. On Earth, however, plasma is the only medium suitable for conducting a controlled thermonuclear synthesis. The idea of an MF thermally insulated plasma (*magnetic bottle*) was proposed in 1950–1951 in USSR and US, for the embodiment of a controlled thermonuclear synthesis. It is based on a thermonuclear reaction in rarefied plasma controlled by a strong MF. Deuterium (D, ^{2}H) and radioactive tritium (T, ^{3}H) are the most promising for the purposes. They react according to the scheme:

$$D + D \rightarrow T + p \; ; \Delta E = -\, 4.0 \text{ MeV} \tag{7.1}$$

$$T + D \rightarrow He + n \; ; \Delta E = -\, 17.6 \text{ MeV} \tag{7.2}$$

It follows from the equations above that it is more promising to use a D/T (1:1) mixture as the thermonuclear fuel, because its energy valuableness is higher than that of pure D. A convincing proof of the prospects of mastering controlled thermonuclear processes is the fact that the energy, evolved in the thermonuclear reaction of D extracted from 1 L of water, equals the energy of combustion of 300 L of petrol.

7.2.5 EXERCISES

E1. Name the characteristic features of plasma. In what does it differ from a conventional gas?

E2. Electrolytes in solution are known to be ionized. In what does plasma differ from dissolved electrolytes?

E3. Since the forces of interaction between highly polar molecules are large, the equation of state of an IG cannot be applied to polar substances (H_2O, NO_2, SO_2, H_2S, etc.). Why does this not relate to isothermal plasma although it consists entirely of charged particles?

E4. A model of the thermonuclear plant T-15 (a doughnut with MF coils), intended for the solution of problems associated with the creation of thermonuclear reactors, was seen at the USSR Exhibition of the Achievements of the National Economy in Moscow. Explain what properties of plasma underlie the design of such a plant.

E5. The temperature of the plasma in a T-10 plant is 2 keV while in a T-15 plant it reaches 10 keV. How many times does the pressure of the plasma in a T-15 plant differ from that in a T-10 one of the concentrations of the particles in the plants is the same?

E6. Calculate the annual consumption of electrical energy by plasmotrons producing HNO_3 with an output of $670{,}000 t \cdot y^{-1}$, proceeding from the fact that a laboratory installation for producing 320 g of NO_x from the air consumes 1 kW•h.

7.3 PHOTOCHEMICAL AND RADIOCHEMICAL REACTIONS

7.3.1 GENERAL

Chain reactions (CRs) are a diverse and widespread phenomenon. They are conditionally divided into several basic types differing in certain features (e.g., combustion, photochemical, polymerization, radiochemical, nuclear reactions). The most important among them are photochemical reactions. The appearance and existence of life on Earth result because of them. *Photochemical* reactions are induced by the action of visible (VIS) and ultraviolet (UV) light, while the branch of physical chemistry studying the reactions is called *photochemistry*. Grotthus (1818) discovered and Draper (1843) formulated the *first law of photochemistry* (Grotthus-Draper law): *photochemical reactions can be induced only by the part of incident light that is absorbed by the reaction system, i.e., is complementary to its color.* Einstein (1912) discovered the *second law of photochemistry* (*law of quantum equivalence*): *every absorbed quantum of light causes the transformation of one molecule.* By a quantum of light is meant its *elementary particle*: a *photon* that is a definite energy pulse, whose value strictly conforms to the frequency of the oscillations or the wavelength of the light containing it:

$$\varepsilon = hv = h\frac{c}{\lambda}$$

(7.3)

where, ε is the energy of a photon, h, Planck's constant (6.626×10^{-34} J·s^{-1}), c, the speed of light in vacuum (3×10^{8} m·s^{-1}), v, the frequency of oscillations and λ, the wavelength of the light. A photochemical reaction consists of two main stages. The initial one includes the primary processes directly induced by light, i.e.:

1. The excitation of molecules leading to the formation of activated particles, i.e., $M + hv \rightarrow M^*$;
2. The dissociation of molecules into radicals, i.e., $B_2 + hv \rightarrow 2B^{\bullet}$;
3. The ionization of the molecules, i.e., $M + hv \rightarrow M^+ + e^-$.

An activated particle exists 10^{-8}s after which it loses its excess energy by radiation [*fluorescence* (FLU)], conversion of it into heat or transferring to another particle when colliding with it (*sensitization*). The substances receiving light quanta and transmitting their energy to other particles are called *sensitizers,* e.g., in the process of the formation of H$^{\bullet}$ radicals in the presence of Hg$_{(v)}$:

$$Hg + hv \rightarrow Hg^* \ ; \ Hg^* + H_2 \rightarrow Hg + 2H^*$$

(7.4)

where, Hg is a sensitizer. When the interaction of the activated particles, formed in the primary processes with other particles of the system, leads to chemical transformations, and one can say that the photochemical process entered its next stage: that of secondary processes, which require no light: they are called *dark processes*. Photochemical reactions are characterized quantitatively by introducing the concept of the quantum yield γ. It equals the number of reacted molecules divided by the number of absorbed photons. If a system receives Φ_e J of light energy a second, i.e., the power of the light is Φ_e W, then the number of photons entering it is $N_{ph} = \Phi_e/h\nu$ s^{-1}. The rate of a photochemical reaction, expressed by the number of molecules that react in one second, is determined by the expression:

$$w = \frac{\gamma\Phi_e}{h\nu} = \frac{\gamma\Phi_e\lambda}{hc} s^{-1}$$

(7.5)

The rate of a photochemical reaction is proportional to the radiant power Φ_e, the length of the light wave λ and the quantum yield γ. Depending on the nature of the secondary processes, the value of γ varies within broad limits, which is why the quantum yield γ is the main characteristic when classifying photochemical reactions. When $\gamma = 1$, reactions are considered to be purely photochemical ones, which are few in number and generally used in laboratory practice for determining the number of absorbed light quanta. Reactions with $\gamma < 1$ generally proceed in solutions or are sensitized, in which cases a part of the absorbed energy is dissipated when the activated particles collide with the solvent molecules, or energy is transferred from the sensitizer molecules to those of the reactants, e.g., the reaction of carbohydrate (CH_2O) synthesis in plants:

$$6CO_2 + 6H_2O + h\nu \rightarrow C_6H_{12}O_6 + 6O_2$$

(7.6)

Eqn. (7.6) is sensitized by chlorophyll and presents a quantum yield of $\gamma \approx 0.1$. Reactions with $\gamma > 1$ relate to CRs. Their quantum yield may present high values, e.g., for the reaction of HCl formation from Cl_2 and H_2, $\gamma \approx 10^5$. Reactions of the type are called *photocatalytic* ones. A feature of photochemical reactions is that their rate does not practically depend on the temperature. The explanation is that the change in the energy of the molecules, when the temperature is increased, is small in comparison with the energy of an absorbed quantum. *Radiochemical reactions* (*radiolysis*) proceed, unlike photochemical ones, under the action of high-energy radiations. The latter are generally a flux of e^-, n^0, p^+, α-particles (He^{2+}), etc., and X and γ-rays. They all result in a greater excitation of the molecules than in photochemical reactions. In other

Plasma, Photo/Radiochemical Reactions, and Relativity Theories 143

respects (mechanism of a process, general laws, etc.), radiochemical reactions are similar to photochemical ones. Photochemical and radiochemical reactions are of a major significance in nature. They are responsible for processes going on with a rise in Gibbs function of the system, and decay in the entropy of the universe, i.e., non-spontaneous processes, which aspect points to the alluring prospect of using the reactions for a number of syntheses. At present, radiochemical reactions are directly used in medicine for treating diseases.

7.3.2 EXERCISE

E1. The decomposition of $KMnO_4$ in an aqueous solution, under the action of light with a wavelength of 322 nm, goes on with a quantum yield of $\gamma = 0.5$. Calculate the rate of decomposition if the radiant power is 0.012 mW. After what time will the concentration of the solution change by 0.001 mol·m^{-3} if the volume of the system is 0.001 m³? *Answer:* $9.72 \times 10^{12} s^{-1}$; 17.2 h.

7.3.3 SOLUTION

S1.
$$w = \frac{\gamma \Phi_e \lambda}{hc} = \frac{0.5 \cdot 1.2 \cdot 10^{-5} \cdot 3.22 \cdot 10^{-7}}{6.626 \cdot 10^{-34} \cdot 3 \cdot 10^{8}} = 9.72 \cdot 10^{12} s^{-1}$$

$$t = \frac{\Delta c}{\dfrac{w}{N_A}} = \frac{10^{-6}}{\dfrac{9.72 \cdot 10^{12}}{6.022 \cdot 10^{23}}} = 61955s = 17.2 h$$

7.4 RELATIVITY THEORIES, PARTICLE PHYSICS, ACCELERATORS, ZOOS, AND MODELS

Aparici proposed questions, answers, and hypotheses on both theories of relativity [7]:

Q1. What is light?
Q2. How does light move?
Q3. What would it occur if one could reach light?
Q4. Was Michelson and Morley experiment either unsuccessful or revolutionary?

Q5. (Einstein). What would it occur if an observer could *reach* a light ray and see it *passed*?

H1. (Einstein). Only way that light movement would not defy logic was that speed were always the same, independently of who would observe it: c = cnt.

H2. (Einstein). This implied that Michelson and Morley experiment would not function.

Q6. What is gravity?

H3. Equivalence principle (Einstein): Gravity is an inertia force and unnecessary; in space far from any stars/freely falling is indistinguishable.

Q7. Corollary: Is the force of gravity necessary?

Q8. However, how? Planet orbits are curves, not straight lines!

H4. Relativistic world (Wheeler): Matter tells space how this must curve; space tells matter how this must move: $R\mu v + (1/2)g\mu vR = (8\pi G/c^4)T\mu v$.

Q9. Does not everybody move at speed $v = c$?

A9. No, in the space-time there are spatial speed and quadri speed; we talked about the former.

Q10. Space can move back but time, cannot; are space and time linked?

A10. Movements mix space/time, which presents something different: entropy (*arrow of time*).

Q11. Where is the European gravitational-wave (GW) detector?

A11. Virgo is in Italy, near Pisa.

Q12. Are there usual applications of the theory of relativity?

A12. Yes, global positioning system (GPS); clocks in satellites put on and must be synchronized.

He proposed questions and answer on particle physics as accelerators, *zoos*, and models.

Q13. (Oppenheimer). Who did order the muon (μ^-)?

Q14. Why the electron (e^-) is stable and the muon (μ^-), not?

A14. Because the electron (e^-) cannot disintegrate.

Q15. Electron (e^-)?

7.5 THE PERFECT THEORY

Ferreira raised the following questions on Einstein's theory of relativity (*cf.* Figure 7.2) [8]:

Q1. How does Einstein's theory of relativity get complicated with the crucial points of the history?
Q2. What is the universe made of?
Q3. How did one find out on what the universe is made of?

FIGURE 7.2 The perfect theory.

7.6 EINSTEIN, THE VACUUM AND THE NOTHINGNESS

De Rújula raised the following questions on Einstein, the vacuum, and the nothingness [9]:

Q1. What are things?
Q2. How simple are they?
Q3. How is the gigantic variety of everything that exists governed by laws that are few and simple?
Q4. The equivalence principle, is it true?
Q5. Is there a giant black hole (BH) of mass $M_{BH} = 4.3 \times 10^6 M_S$ ($M_S = M_{Sun}$) in the Milky Way center?
Q6. Cosmos?
Q7. How do things work?

Q8. Cosmological constant: Einstein's biggest mistake?

He provided the following conclusions (Cs):

C1. The vacuum is not the nothingness; it is a substance $\varphi \neq 0$.

C2. If the vacuum is a substance, one can shake it; its vibrations are elementary particles: the Higgs.

C3. The SMs of micro and macrocosmos burst.

C4. Einstein invented GTR and QM wave-particle photon but nobody knows how to unify them.

C5. In Einstein's blackboard, there is a lot to fill.

7.7 FINAL REMARKS

From the present results and discussion, the following final remarks can be drawn:

1. The quantum yield is the main characteristic when classifying photochemical reactions, because its value varies within broad limits.

2. A feature of photochemical reactions is that their rate does not practically depend on the temperature.

3. Photochemical and radiochemical reactions are responsible for processes going on with a rise in Gibbs function of the system, and a decay in the entropy of the universe, i.e., non-spontaneous processes, which aspect points to the alluring prospect of using the reactions for a number of syntheses.

4. The vacuum is not the nothingness; it is a substance $\varphi \neq 0$.

5. If the vacuum is a substance, one can shake it; its vibrations are elementary particles: the Higgs.

6. The SMs of micro and macrocosmos burst.

7. Einstein invented GTR and QM wave-particle photon, but nobody knows how to unify them.

8. In Einstein's blackboard, there is a lot to fill.

ACKNOWLEDGMENTS

The authors thank support from Fundacion Universidad Catolica de Valencia San Vicente Martir (Project No. 2019-217-001UCV).

KEYWORDS

- **global positioning system**
- **hydrocarbons**
- **particle physics**
- **plasma application**
- **plasma generation**
- **quantum mechanics**
- **vacuum**

REFERENCES

1. Torrens, F., & Castellano, G., (2019). Nuclear fusion and the American nuclear cover-up in Spain: Palomares disaster (1966). In: Haghi, R., & Torrens, F., (eds.), *Engineering Technology and Industrial Chemistry with Applications* (pp. 297–308.). Apple Academic-CRC: Waretown, NJ.

2. Torrens, F., & Castellano, G., (2020). Manhattan project, atoms for peace, nuclear weapons, and accidents. In: Pogliani, L., Torrens, F., & Haghi, A. K., (eds.), *Molecular Chemistry and Biomolecular Engineering: Integrating Theory and Research with Practice* (pp. 215–233). Apple Academic-CRC: Waretown, NJ.

3. Torrens, F., & Castellano, G., (2021). Nuclear science and technology. In: Pogliani, L., Ameta, S. C., & Haghi, A. K., (eds.), *Chemistry and Industrial Techniques for Chemical Engineers.* Apple Academic-CRC: Waretown, NJ, in press. pp. 311–330.

4. Torrens, F., & Castellano, G. History, concept, method, and didactics of atomic/nuclear physics. In: Esteso, M. A., Ribeiro, A. C. F., & Haghi, A. K., (eds.), *Physical Chemistry and Its Interdisciplinary Applications.* Apple Academic–CRC: Waretown, NJ, in press.

5. Torrens, F., & Castellano, G. Gravitational waves, messengers of the universe, and cosmology. In: Esteso, M. A., Ribeiro, A. C. F., & Haghi, A. K., (eds.), *Chemistry and Chemical Engineering for Sustainable Development: Best Practices and Research Directions.* Apple Academic-CRC: Waretown, NJ, in press.

6. Akhmetov, B., Novichenko, Y., & Chapurin, V., (1989). *Physical and Colloid Chemistry.* Mir: Moscow, U.S.S.R.

7. Aparici, A., (2019). *Personal Communication.*

8. Ferreira, P., (2019). *Personal Communication.*

9. De Rújula, Á., (2019). *Personal Communication.*

CHAPTER 8

Mesoporous, Graphene Composite, Li-Battery, Topology, and Periodicity

FRANCISCO TORRENS[1] and GLORIA CASTELLANO[2]

[1]*Institute for Molecular Science, University of Valencia, PO Box 22085, E–46071 Valencia, Spain, E-mail: torrens@uv.es.*

[2]*Department of Experimental Sciences and Mathematics, Faculty of Veterinary and Experimental Sciences, Valencia Catholic University Saint Vincent Martyr, Guillem de Castro-94, E–46001 Valencia, Spain*

ABSTRACT

The history of electronics began with germanium, passed to silicon, and finally to carbon. The periodic table of the elements (PTE) of the history of electronics was reported. Perhaps mesoporous silica-based materials will be the future of electronics. Topological quantum chemistry (TQC) is a predictive theory of topological hands. Taking the ideas to their logical conclusion, a paradigm appears which applies not only to topological insulators but also to semimetals and band theory. The synthesis of symmetry and topology of localized orbitals and Bloch wave functions must enable a full understanding of noninteracting solids. The emphasis on the symmetry of localized orbitals opens up incorporating magnetic groups or interactions in the theory of topological materials. Predicted *d/f*-electron compounds must be analyzed by more accurate dynamical *mean-field theory* codes, which properly take into account interactions and can potentially give many new interacting materials with strong topology. Mimics of two-dimensional materials are possible in practice with the products outperforming the bulk materials, while control over stability can be governed by immobilization on solid surfaces. It was investigated how best to integrate experiment, computation, and theory. Science is seen from outside as experimental but another theoretical science exists, and classification is a part of theoretical science.

8.1 INTRODUCTION

Setting the scene: mesoporous materials from synthesis to applications, graphene (GR) nanocomposites (NCs), lithium batteries, the periodic table of topological materials, sustainability, the importance of the history of science and the solution of the knowledge of the factors hindering its translation to teaching practice. Two-dimensional (2D) materials present interesting properties because of their two-dimensionality, e.g., hardness, flexibility, and a large number of favorable electronic and thermal characteristics. The history of electronics began with Ge, then passed to Si and, finally, to C. The periodic table of the elements (PTE) of the history of electronics was reported. Perhaps mesoporous silica ($mSiO_2$)-based materials will be the future of electronics. Topological quantum chemistry (TQC) results a predictive theory of topological hands. Taking the ideas to their logical conclusion, a paradigm appears which applies not only to topological insulators but also to semimetals and band theory. The synthesis of symmetry and topology of localized orbitals and Bloch wave functions must enable a full understanding of noninteracting solids. The emphasis on the symmetry of localized orbitals opens up incorporating magnetic groups or interactions in the theory of topological materials. In physics and probability theory, *mean-field* [MFT, or *self-consistent field* (SCF), e.g., Hartree-Fock (HF)] *theory* studies the behavior of high-dimensional random (stochastic) models by studying a simpler model that approximates the original by averaging over degrees of freedom. Predicted *d/f*-electron compounds must be analyzed by more accurate dynamical MFT codes, which properly take into account interactions; the codes can potentially give many new interacting materials with strong topology. Mimics of 2D materials are possible in practice with the products outperforming the bulk materials, while control over stability can be governed by immobilization on solid surfaces. It was investigated how best to integrate experiment, computation, and theory. Science is seen from outside as experimental but another theoretical science exists, and classification is a part of theoretical science. The importance of the history of science follows. Science is a human and, so, social activity. The history (of science) allows a better understanding of complex processes. It makes it easier to reflect critically on the present.

The $mSiO_2$-based materials for electronics-oriented applications were reviewed [1]. The analysis of organophosphates in Li^+ battery electrolytes was reported by hydrophilic interaction liquid chromatography (HILIC)-electrospray ionization (ESI)-mass spectrometry (MS) [2].

In earlier publications, it was informed the effects of type, size, and elliptical deformation on molecular polarizabilities of model single-wall carbon nanotubes (SWNTs) from atomic increments [3–6], SWNTs periodic properties and table based on the chiral vector [7, 8], calculations on SWNTs solvents, co-solvents, *cyclo*pyranoses [9–12] and organic-solvent dispersions [13, 14], packing effect on cluster nature of SWNTs solvation features [15], information entropy analysis [16], cluster origin of SWNTs transfer phenomena [17], asymptotic analysis of coagulation-fragmentation equations of SWNT clusters [18], properties of fullerite and symmetric C-forms, similarity laws [19], fullerite crystal thermodynamic characteristics, law of corresponding states [20], cluster nature of nanohorns (SWNHs) solvent features [21], SWNTs (co-)solvent selection, *best* solvents, acids, superacids, host-guest inclusion complexes [22], $C/BC_2N/BN$ fullerenes/SWNTs/nanocones (SWNCs)/SWNHs/buds (SWNBs)/GRs cluster solvation models in organic solvents [23–29], elementary polarizability of Sc/*fullerene*/GR aggregates, di/GR-cation interactions [30], the world of materials from layered ones to conductive 2D metal-organic frameworks (MOFs) [31–35], porous materials (PMs), MOFs, conduction [36], magnetism, polyoxometalates (POM), layered materials and GR [37]. The purpose of the present report is to review mesoporous materials from synthesis to applications, GR NCs, Li batteries, the periodic table of topological materials, sustainability, the importance of the history of science and the solution of the knowledge of the factors hindering its translation to teaching practice. The aim of this work is to initiate a debate by suggesting a number of questions (Q), which can arise when addressing subjects of the periodic table of topological materials. It was provided, when possible, hypotheses (H) on these materials and sustainability.

8.2 MESOPOROUS MATERIALS: FROM SYNTHESIS TO APPLICATIONS

The $mSiO_2$ results inorganic materials, which are formed by the condensation of Na^+ silicate or Si alkoxides around an ordered surfactant used as template, which synthesis and porous structure depend on several parameters (e.g., Si source, morphology of the surfactant, pH conditions, ionic strength, aging temperature and time), which synthetic conditions directly affect the structure, specific surface area (SSA), pore size and volume or wall thickness [38]. The $mSiO_2$ presented a great impact in 1990s, when Mobil Oil Corporation designed Mobil composition material (MCM) with different morphological characteristics (in 1992). They synthesized MCM-41, -48, and -50 mesophases, which showed hexagonal, cubic, and lamellar mesostructures, respectively. The porous SiO_2 were synthesized *via* alkyl NH_4^+ salts as the structure directing agent, and Na^+ silicate or tetraethylorthosilicate (TEOS) in basic medium as Si source. The pore size did not exceed 4.0 nm and the wall thickness was below 2.0 nm. Zhao et al. (1998) first reported an ordered mesoporous material [*Santa Barbara Amorphous* (SBA) *mesoporous molecular sieve*]. The SBA-15 with hexagonal ordering and -16 with lamellar structure were widely studied later. The SBAs are synthesized *via* non-ionic surfactants derived from poly(propylene oxide) (PPO) and poly(ethylene oxide) (PEO), as pore directing agents in acidic medium. The main advantage of the porous SiO_2 in comparison to MCM results related to the larger pore thickness, which provides them with greater thermochemical stability. The pore size of SBA was modulated in 2–50 nm, expanding the number of applications. A group at Michigan State University synthesized other porous SiO_2 (MSU) *via* non-ionic surfactants at neutral p*H*, obtaining a pore diameter in 2.1–8.0 nm and a wall thickness in 1.5–4.0 nm; however, MSUs display less organized structures than MCM/SBAs. Since 1990s, the pore size resulted modulated *vs.* the application. The textural properties were modified controlling the synthetic conditions (aging time and temperature), by the insertion of swelling organic molecules [aliphatic hydrocarbons (HCs), aromatics, alkylamines], adjusting of template, by the insertion of heteroatoms in the porous SiO_2 framework or changing the removal conditions of the template.

8.3 GRAPHENE (GR) NANOCOMPOSITES (NCS)

Integrating GR with other nanomaterials (NMs) created a variety of NCs with extraordinary chemical, optical, mechanical, and electrical properties, which combined the characteristics of their components to achieve structural stability and multifunctionality with proper design of the interfacial interactions [39]. The field of GR NCs developed rapidly over 2000's with many applications in a number of fields (e.g., chemical/biomedical sensing, tissue engineering, bioimaging).

8.4 LITHIUM BATTERIES

After Bard et al. the term *Li batteries* refers to a family of different devices presenting in common a Li metal anode whose discharge reaction is: Li

$\rightarrow Li^+ + e^-$ [40]. The Li batteries found application in portable electronic devices owing to their high working voltage (4V), high energy density, excellent cyclability, flat discharge characteristics, and long shelf life (10y). The Li was initially used as a cathode material and was later replaced (because of its reactivity) by Li-C intercalation materials. Cathode materials include lithiated transition metal oxides, e.g., $LiCoO_2$, $LiNiO_2$, $LiMn_2O_4$, and related compounds, apart from which, different PMs were studied for Li batteries. Abraham et al. Desilvestro and Haas, and Delmas et al. described V_2O_5 and vanadates (LiV_3O_8, V_6O_{13}); recent advances include sol-gel-Cr-modified V_6O_{13}. Structurally related materials are MoO_2, MoS_2, and $MnMoO_4$, which possess high cycling capability. The second group of widely studied materials comprises MnO_2 and related compounds, Fe phosphate-based materials, and Fe sulfides (FeS, FeS_2). The Li batteries operate over a wide temperature (-40 to $-70°C$), the most common solvents being polar organic liquids, e.g., acetonitrile, propylene carbonate, methyl formate, etc. Thionyl ($SOCl_2$) and sulfuryl (SO_2Cl_2) chlorides are inorganic liquids that are used as the solvent and active cathode material in Li batteries. The $LiClO_4$, $LiBF_4$, and $LiAsF_6$ are the most common salts for supporting electrolyte dosage. Developments include using solid electrolytes, i.e., polyethers, thin films (TFs) of conducting ceramic materials, etc. The electrochemical performance of cathodes depends significantly on their behavior with regard to Li diffusion, in turn, depending on particle size, surface morphology, homogeneity, and porosity. Porous composite cathodes are regarded as two contiguous interwoven nets. The cathode-active material net consists of electronically conducting particles, whereas the second net is a polymer electrolyte that fills the pores between the active cathode material particles. The Li^+ participating in the cathode reaction and originating from the Li electrode migrate *via* the composite into the polymer electrolyte-filled pores of the composite cathode to react with the cathode-active material. The mechanism of charge-discharge may be affected by ion transport in the electrolyte net and *redox* insertion kinetics of the cathode-active material. General requirements follow: (1) electrical contact must be ensured between the particles of the cathode-active material; (2) the electrolyte must completely fill the pores to prevent the separation of cathode-active particles and provide ion conductivity between them; (3) the mobility of ions in the polymer-electrolyte phase must be comparable to that in the cathode-active material phase. The cycle life is defined as the number of charge/discharge cycles that a battery goes through, under given conditions, before it reaches predefined minimum performance limits. The cycle of life depends on charge and discharge rates and cutoff limits, depth of discharge, self-discharge rate, and temperature of operation. The cyclic voltamperometer (CV) of a $LiMn_2O_4$ electrode was measured on a cell with Li foil for the reference and auxiliary electrodes in ethylene plus dimethyl carbonates solution of $LiAsF_6$ (1 mol·L^{-1}). The pair of peaks at larger potential corresponds to the deintercalation/intercalation of Li in $0.0 \leq x \leq 0.5$ for $Li_xMn_2O_4$, whereas the pair of peaks at lower potentials is attributable to the process for $0.5 \leq x \leq 1.0$, both accompanied by reversible Mn^{IV}/Mn^{III} redox reactions. After Xia and Yoshio, the later electrochemical process corresponds to the removal/addition of Li^+ from/into half of the tetrahedral (T_d) sites in which the Li intercalation occurs. The former couple is then attributed to the process at the other T_d sites where Li intercalations do not occur. Electrochemical impedance spectroscopy (EIS) for a $LiMn_2O_4$ electrode at 20°C was measured, whereas the equivalent circuit for fitting experimental impedance data was depicted. Nyquist plot is described in terms of two superimposed semicircles. Because the semicircles are depressed with their centers below the real axis, constant phase elements (CPE) are introduced in the circuit. The R_1 and Q_1 represent the resistance of the surface film on $LiMn_2O_4$ and the CPE representing the film capacitance, both corresponding to the high-frequency (HF) region of the impedance spectrum. The R_2 and Q_2 denote, respectively, the resistance and CPE corresponding to the electrochemical intercalation/deintercalation process and double-layer capacitance. The R_2 is equivalent to the charge-transfer resistance R_{ct} of Mn^{4+}/Mn^{3+} redox process, which magnitude exhibits an Arrhenius-type variation *vs.* temperature, so that the activation energy E_a of the electrochemical process is obtained from the slope of $\ln(R_{ct}/T)$ *vs.* $1/T$ linear relationship:

$$\frac{\Delta\left[\ln\left(R_{ct}/T\right)\right]}{\Delta\left(1/T\right)} = \frac{E_a}{R} \tag{8.1}$$

In the case of pyrite batteries, Strauss et al. proposed a multistep mechanism for the charge-discharge of pyrite in polymer electrolytes in the temperature 90–135°C. In the first cycle, the reduction of pyrite occurs in two steps:

$$FeS_2 + 2Li^+ + 2e^- \rightarrow Li_2FeS_2 \tag{8.2}$$

$$Li_2FeS_2 + 2Li^+ + 2e^- \rightarrow Fe + 2Li_2S \tag{8.3}$$

The second cycle charge/discharge curve for Li/composite polymer electrolyte/FeS_2 cell was measured. Region-b in the voltage/capacity curve corresponds to the reversible electrochemical reaction:

$$Fe + 2Li_2S \rightarrow Li_2FeS_2 + 2Li^+ + 2e^- \tag{8.4}$$

Whereas region-a is associated with the reversible insertion/deinsertion of Li in Li_2FeS_2 host, complicated by phase change. The phase evolution of the system was schematized. In the first step, Fe metal is oxidized to Fe^{2+} at Fe/Li_2S interface (I) and a Li_2FeS_2 layer is formed in that interface:

$$(Fe)_I \rightarrow (Fe^{2+})_I + 2e^- \tag{8.5}$$

$$(Fe^{2+})_I + 2(Li_2S_I) \rightarrow (Li_2FeS_2)_I + 2Li^+ \tag{8.6}$$

where, Li^+ are released, whereas two interfaces, Fe/Li_2FeS_2 (II) and Li_2FeS_2/Li_2S (III), are generated. The Fe^{2+} formed at interface II migrate and diffuse *via* lattice defects in Li_2FeS_2 phase until they reach interface III, where they react with Li_2S:

$$(Fe)_{II} \rightarrow (Fe^{2+})_{II} + 2e^- \tag{8.7}$$

$$(Fe)_{II} \rightarrow (Fe^{2+})_{II} \tag{8.8}$$

$$(Fe^{2+})_{III} + 2(Li_2S)_{III} \rightarrow (Li_2FeS_2)_{III} + 2Li^+ \tag{8.9}$$

The step continues until all Fe particles electrically connected to the current collector react with Li_2S particles, which produces a thickening of the intermediate layer of Li_2FeS_2. When a given threshold thickness is reached, a potential jump appears because of the transport limitations of Fe^{2+} in Li_2FeS_2. Final steps lead to the formation of a layer of $Li_{2-x}FeS_2$, in which Fe^{2+} is oxidized to Fe^{3+} coupled to deinsertion of Li^+, and a charge in Fe occupancy at an advanced stage of the reaction. The charging and discharging processes of Li batteries involve the transfer of Li^+ from one ion insertion electrode to another, which process is regarded as a topotactic intercalation reaction of Li^+ into interstitial sites in the crystalline host matrices, eventually accompanied by first-order phase transitions (PTs). Electrochemical detection of such PTs is conditioned by slow solid-state diffusion of ions into the solid matrix and uncompensated Ohmic drops. The PT is associated with the two-peak response in CVs, corresponding to a composite graphite electrode immersed into $LiPF_6$ (ethylene plus dimethyl carbonates) electrolyte. The electrochemical conversion of $LiC_{12} \rightarrow LiC_6$ was depicted. The most singular picture of PT/insertion process is provided by chronoamperometric (CA) curves, corresponding to the intermittent titration of graphite electrode across the major intercalation peak for $LiC_{12} \rightarrow LiC_6$ transition, potentiostatic titration in 0.085–0.080 V, where the low-mass film of powdery composite electrodes in a coin-type cell configuration, in contact with a small amount of $LiPF_6$ in ethylene plus dimethyl carbonates electrolyte, was used to improve data resolution. Five regions, labeled I–V, were distinguished in CA curve: At short times (region I), double-layer charging is accompanied by saturation of the initial phase, LiC_{12}, with Li^+, where nuclei with a subcritical size are formed, further undergoing co-alescence to form larger size, supercritical nuclei, a well-known phenomenon in metal deposition electrochemistry. In region II, the current rises almost linearly with time, a process that is associated with the growth of supercritical nuclei, overlapping with each other and finally forming a continuous layer. The co-alescence of nuclei of the new phase (region III) determines the end of the growth of the total surface area. The current decays *via* region IV, where the boundary between the two co-existing phases advances towards the electrode bulk. Finally, the current decays monotonically when the electrode reaches equilibrium with Li^+ in the solution at the applied potential (region V). Two alternative models interpret the current decay in region IV: a quasi-Cottrellian approach and a progressive boundary motion model. In the former, it is assumed that the quotient, between the charge passed at time t' after the maximum current $Q_{t'}$ and total charge passed at infinite time Q_∞, is proportional to the square root of time:

$$\frac{Q_{t'}}{Q_\infty} = 4\left(\frac{Dt'}{\pi a}\right)^{1/2}$$

(8.10)

Assuming that diffusion of Li^+ (with diffusion coefficient D) occurs from two opposite edges of a particle of size a. In the latter model, $Q_{t'}/Q_\infty$ ratio becomes:

$$\frac{Q_{t'}}{Q_\infty} = 2\gamma\left(D_{phase}t\right)^{1/2}$$

(8.11)

where, D_{phase} is the diffusion coefficient for the propagating phase and γ, a numerical coefficient whose value approaches 0.52. Diffusion influences notably ESI responses for the kind of system. When conditions of semi-infinite diffusion hold, Nyquist plot comprises a semicircle followed by a linear branch. When the thickness of the stagnant layer is finite (trans-missive finite diffusion), two semicircles are obtained, while if transmission takes place at a limited distance (reflective finite diffusion), two consecutive linear regions follow to HF semicircle.

8.5 PERIODIC TABLE OF TOPOLOGICAL MATERIALS

García Vergniory proposed Q/H on the periodic table of topological materials (*cf.* Figure 8.1) [41]:

Q1. What does it make topological materials unique?
H1. Topological quantum chemistry (TQC) [42].
Q2. How many topological materials do them exist in nature?
Q3. What are they?
Q4. What is their abundance?
Q5. How can energy bands in solid be connected throughout Brillouin zone (BZ) to obtain all realizable band structures in all non-magnetic space-groups?
Q6. How do real-space orbitals in a material determine the symmetry character of electronic bands?
Q7. How does TQC apply to GR with spin-orbit coupling (SOC)?
Q8. What have they done?
Q9. Which materials do present topological properties?
Q10. How, *via* TQC, are we able to catalog all topological materials that can possibly exist in nature [43]?
Q11. Why topology?
H2. Euler's polyhedron formula: $V + F - E = 2$ (V: vertices, F: faces, E: edges).
H3. Closed surfaces, Gauss-Bonnet: $\int_M \kappa dA = 2\pi\chi = 2\pi(2-2\,g)$, where g is the number of holes.
H4. Open surfaces, Barry curvature: $\sigma_h = ne^2/h$.
Q12. What is a topological insulator?
H5. Block theorem: theory of electronic band structure.
Q13. How do we classify topological insulators?
Q14. Topology?
Q15. Why are elementary band representations important?
H6. Periodic table of topological materials [S, Ge].

FIGURE 8.1 Modeled interference patterns formed by quantum waves in topological material.

Source: A. Yazdani/SPL.

She provided the following conclusion (C):

C1. TQC: predictive theory of topological hands.

8.6 SUSTAINABILITY

Ganin proposed the following hypotheses on sustainability [44]:

H1. Challenge: Renewable energy (RE) is *intermittent*→ energy storage.
H2. Recycling RE: domestic storage.
H3. Looking ahead… rebound effect, hedge *vs.* unknown unknowns, giving back control to achieve completely sustainable future.

8.7 DISCUSSION

The 2D materials present interesting properties because of their two-dimensionality, e.g., hardness, flexibility, and a large number of favorable electronic and thermal characteristics.

8.8 FINAL REMARKS

From the present results and discussion, the following final remarks can be drawn.

1. The history of electronics began with germanium, then passed to silicon and, finally, to carbon (with oxidation states (OSs) of 4 and -4). The PTE of the history of electronics was reported. Perhaps mesoporous silica-based materials will be the future of electronics. TQC is a predictive theory of topological hands.

2. Taking the ideas to their logical conclusion, a new paradigm appears which applies not only to topological insulators but also to semimetals and band theory, generally. The synthesis of symmetry and topology of localized orbitals and Bloch wave functions must enable a full understanding of noninteracting solids. The emphasis on the symmetry of localized orbitals opens up an avenue to incorporating magnetic groups or interactions in the theory of topological materials.

3. The d/f-electron compounds that were predicted should be analyzed by more accurate dynamical mean-field theory codes, which properly take into account interactions; the codes can potentially give many new interacting materials with strong topology.

4. Mimics of two-dimensional materials are possible in practice with the products outperforming the bulk materials, while control over stability can be governed by immobilization on solid surfaces. It was investigated how best to integrate experiment, computation, and theory. Science is seen from outside as experimental but another theoretical science exists, and classification is a part of theoretical science.

5. The importance of the history of science follows. Science is a human and, so, social activity. The history (of science) allows a better understanding of complex processes. It makes it easier to reflect critically on the present. It remains to be solved the knowledge of the factors hindering the translation of research in the history of science to teaching practice. As educators of the professionals of tomorrow, professors have the ethical duty of providing an integral training, where teachers should include environment protection.

ACKNOWLEDGMENTS

The authors thank support from Fundacion Universidad Catolica de Valencia San Vicente Martir (Project No. 2019-217-001UCV).

KEYWORDS

- **electrode**
- **electrospray ionization**
- **functionalized silica**
- **low-*k* dielectric**
- **mesoporous silica material**
- **molecular electronics**
- **supercapacitor**

REFERENCES

1. Laskowski, L., Laskowska, M., Vila, N., Schabikowski, M., & Walcarius, A., (2019). Mesoporous silica-based materials for electronics-oriented applications. *Molecules, 24,* 2395–1–31.
2. Henschel, J., Winter, M., Nowak, S., & Jiang, W., (2017). Analysis of organophosphates in lithium ion battery electrolytes by HILIC-ESI-MS. *LC·GC Eur., 30*(The Applications Book), 691–692.
3. Torrens, F., (2003). Effect of elliptical deformation on molecular polarizabilities of model carbon nanotubes from atomic increments. *J. Nanosci. Nanotech., 3,* 313–318.
4. Torrens, F., (2004). Effect of size and deformation on polarizabilities of carbon nanotubes from atomic increments. *Future Generation Comput. Syst., 20,* 763–772.
5. Torrens, F., (2004). Effect of type, size, and deformation on polarizability of carbon nanotubes from atomic increments. *Nanotechnology, 15,* S259–S264.
6. Torrens, F., (2006). Corrigendum: Effect of type, size, and deformation on polarizability of carbon nanotubes from atomic increments. *Nanotechnology, 17,* 1541–1541.
7. Torrens, F., (2004). Periodic table of carbon nanotubes based on the chiral vector. *Internet Electron. J. Mol. Des., 3,* 514–527.
8. Torrens, F., (2005). Periodic properties of carbon nanotubes based on the chiral vector. *Internet Electron. J. Mol. Des., 4,* 59–81.
9. Torrens, F., (2005). Calculations on *cyclo*-pyranoses as co-solvents of single-wall carbon nanotubes. *Mol. Simul., 31,* 107–114.
10. Torrens, F., (2005). Calculations on solvents and co-solvents of single-wall carbon nanotubes: *Cyclo*-pyranoses. *J. Mol. Struct. (Theochem.), 757,* 183–191.

11. Torrens, F., (2005). Calculations on solvents and co-solvents of single-wall carbon nanotubes: *Cyclo*-pyranoses. *Nanotechnology, 16*, S181–S189.

12. Torrens, F., (2005). Some calculations on single-wall carbon nanotubes. *Probl. Nonlin. Anal. Eng. Syst., 11*(2), 1–16.

13. Torrens, F., (2006). Calculations of organic-solvent dispersions of single-wall carbon nanotubes. *Int. J. Quantum Chem., 106*, 712–718.

14. Torrens, F., & Castellano, G., (2005). Cluster origin of the solubility of single-wall carbon nanotubes. *Comput. Lett., 1*, 331–336.

15. Torrens, F., & Castellano, G., (2007). Cluster nature of the solvation features of single-wall carbon nanotubes. *Curr. Res. Nanotech., 1*, 1–29.

16. Torrens, F., & Castellano, G., (2007). Effect of packing on the cluster nature of C nanotubes: An information entropy analysis. *Microelectron. J., 38*, 1109–1122.

17. Torrens, F., & Castellano, G., (2007). Cluster origin of the transfer phenomena of single-wall carbon nanotubes. *J. Comput. Theor. Nanosci., 4*, 588–603.

18. Torrens, F., & Castellano, G., (2007). Asymptotic analysis of coagulation-fragmentation equations of carbon nanotube clusters. *Nanoscale Res. Lett., 2*, 337–349.

19. Torrens, F., & Castellano, G., (2008). Properties of fullerite and other symmetric carbon forms: Similarity laws. *Symmetry Cult. Sci., 19*, 341–370.

20. Torrens, F., & Castellano, G., (2010). Fullerite crystal thermodynamic characteristics and the law of corresponding states. *J. Nanosci. Nanotechn., 10*, 1208–1222.

21. Torrens, F., & Castellano, G., (2010). Cluster nature of the solvent features of single-wall carbon nanohorns. *Int. J. Quantum Chem., 110*, 563–570.

22. Torrens, F., & Castellano, G., (2011). (Co-)solvent selection for single-wall carbon nanotubes: Best solvents, acids, superacids and guest-host inclusion complexes. *Nanoscale, 3*, 2494–2510.

23. Torrens, F., & Castellano, G., (2012). *Bundlet* model for single-wall carbon nanotubes, nanocones and nanohorns. *Int. J. Chemoinf. Chem. Eng., 2*(1), 48–98.

24. Torrens, F., & Castellano, G., (2013). Solvent features of cluster single-wall C, BC_2N and BN nanotubes, cones, and horns. *Microelectron. Eng., 108*, 127–133.

25. Torrens, F., & Castellano, G., (2013). Corrigendum to: Solvent features of cluster single-wall C, BC_2N and BN nanotubes, cones, and horns. *Microelectron. Eng., 112*, 168–168.

26. Torrens, F., & Castellano, G., (2013). *Bundlet* model of single-wall carbon, BC_2N and BN nanotubes, cones, and horns in organic solvents. *J. Nanomater. Mol. Nanotech., 2*, 1000107–1–9.

27. Torrens, F., & Castellano, G., (2013). C-nanostructures cluster models in organic solvents: Fullerenes, tubes, buds and graphenes. *J. Chem. Chem. Eng., 7*, 1026–1035.

28. Torrens, F., & Castellano, G., (2014). Cluster solvation models of carbon nanostructures: Extension to fullerenes tubes and buds. *J. Mol. Model., 20*, 2263–1–9.

29. Torrens, F., & Castellano, G., (2014). Cluster model expanded to C-nanostructures: Fullerenes, tubes, graphenes and their buds. *Austin J. Nanomed. Nanotech., 2*(2), 7–1–7.

30. Torrens, F., & Castellano, G., (2013). Elementary polarizability of Sc/fullerene/graphene aggregates and di/graphene-cation interactions. *J. Nanomater. Mol. Nanotech., S1*, 001–1–8.

31. Torrens, F., & Castellano, G., (2018). Conductive layered metal-organic frameworks: A chemistry problem. *Int. J. Phys. Study Res., 1*(2), 42–42.

32. Torrens, F., & Castellano, G., (2019). Conductive two-dimensional nanomaterials: Metal-organic frameworks. *J. Appl. Phys. Nanotech., 2*(2), 52–52.
33. Torrens, Z. F., (2019). From layered materials to bidimensional metal-organic frameworks. *Nereis, 2019*(11), 65–80.
34. Torrens, F., & Castellano, G., (2018) Conductive layered metal-organic frameworks: A chemistry problem. *Int. J. Phys. Study Res., 1(2),* 42–42.
35. Torrens, F., & Castellano, G. World of conductive two-dimensional metal–organic frameworks. In: Yaser, A. Z., Khullar, P., & Haghi, A. K., (eds.), *Green Materials and Environmental Chemistry: New Production Technologies, Unique Properties, and Applications.* Apple Academic-CRC: Waretown, NJ, in press.
36. Torrens, F., & Castellano, G. Porous materials, metal-organic frameworks and conduction. In: Esteso, M. A., Ribeiro, A. C. F., & Haghi, A. K., (eds.), *Chemistry and Chemical Engineering for Sustainable Development: Best Practices and Research Directions.* Apple Academic-CRC: Waretown, NJ, in press.
37. Torrens, F., & Castellano, G. Magnetism, polyoxometalates, layered materials and graphene. In: Haghi, A. K., (ed.), In: Kulkarni, S., Rawat, N. K., & Haghi, A. K. (eds.), *Green Chemistry and Green Engineering: Processing, Technologies, Properties, and Applications.* Apple Academic–CRC: Waretown, NJ, in press.
38. Cecilia, J. A., Moreno, T. R., & Retuerto, M. M., (2019). Mesoporous materials: From synthesis to applications. *Int. J. Mol. Sci., 20,* 3213–1–4.
39. Zhao, X., & Yang, M., (2019). Graphene nanocomposites. *Molecules, 24,* 2440–1–2.
40. Doménech-Carbó, A., (2010). *Electrochemistry of Porous Materials.* CRC: Boca Raton, FL.
41. García, V. M., (2019). *Personal Communication.*
42. Bradlyn, B., Elcoro, L., Cano, J., Vergniory, M. G., Wang, Z., Felser, C., Aroyo, M. I., & Bernevig, B. A., (2017). Topological quantum chemistry. *Nature (London), 547,* 298–305.
43. Vergniory, M. G., Elcoro, L., Felser, C., Regnault, N., Bernevig, B. A., & Wang, Z., (2019). A complete catalog of high-quality topological materials. *Nature (London), 566,* 480–485.
44. Ganin, A., (2019). *Personal Communication.*

CHAPTER 9

Boom for Circular Economy and Creativity: Chemistry to Improve Life

FRANCISCO TORRENS[1] and GLORIA CASTELLANO[2]

[1]*Institute for Molecular Science, University of Valencia, PO Box 22085, E–46071 Valencia, Spain, E-mail: torrens@uv.es*

[2]*Department of Experimental Sciences and Mathematics, Faculty of Veterinary and Experimental Sciences, Valencia Catholic University Saint Vincent Martyr, Guillem de Castro-94, E–46001 Valencia, Spain*

ABSTRACT

Creativity in science is the ability to get oneself into a fine mess. Chemophobia false myth: All that *smells* chemistry is bad. Positive meaning: *Between two persons, chemistry exists*. Solubility old rule: *Like dissolves like*. Research is to raise oneself questions and try to answer them. Technicians are the *why* but farmers are the *how*. New ethos: Spain–US agreement. Better things for better living *via* chemistry. Not all fascist science is bad but excellent science requires social freedom. Practically, no historical rhetoric exists and, in Spanish-Civil-War Generation, totals rupture. Ingenuity and particular nature of chemistry for armor-plating itself. Tacit knowledge: intangible. Operation Paperclip: many German scientists recruited in post-Nazi Germany and taken to the US for government employment. Historical processes exist that require a long time. If I were a pupil of baccalaureate, I would ask my center that taught me chemistry well. Chemistry is related to two other sciences: biomedicine and materials science. Chemistry should be taught in a laboratory because it is an experimental science. Here we have the data, what conclusions can we take out from them? Here we have the conclusions, what data can we explain with them?

9.1 INTRODUCTION

Setting the scene: boom time for the *circular economy* (CE), creativity, whether it is all chemistry, chemistry exclusiveness to improve people's life, scientific research to improve people's life, an investment of future, chemical science for everybody, chemistry, energy challenges of the 21st century, the present and future of plant safe, quantifying amorphous contents in milled powders with calorimetry, safety storage cabinets, the control of the safety of fume cupboards, catalysts looking after the environment, graphene (GR) science and technology (*scitech*), whether expectations are coming true, from petroleum to sunlight, solar-fuels photocatalytic production (PCP), the power of chemistry in Franco's Spain, II FUSCHROM Workshop on *Separation Strategies in Chromatography*, microwaves (MWs) as a fundamental tool in the analytical laboratory, the Earth at geologist's eye, contents as resource for encouraging interest to chemistry study, immobilized catalytic systems for asymmetric flow processes, three in one, three viruses and one micromurder, Zika, Chincungunya, and dengue, chemistry of the outer space, seeing the negative of the sky, the intergalactic medium and preparing students for practical sessions *via* laboratory simulation software.

In earlier publications, a comparison was made between beer bars cafés (19th century) [1–5], beer, all a science, alcohol, analysis, quality testing at multiple brewery stages [6], beer molecules, sensory, bioproperties, and street binge drinking [7].

The aim of the present report is to review boom time for CE, creativity, whether it is all chemistry, chemistry exclusiveness to improve people's life, scientific research to improve people's life, an investment of future, chemical science for everybody, chemistry, energy challenges of the 21st century, the present and future of plant safe, quantifying amorphous contents in milled powders with calorimetry, safety storage cabinets, the control of the safety of fume cupboards, catalysts looking after the environment, GR *scitech*, whether expectations are coming true, from petroleum to sunlight, solar-fuels PCP, the power of chemistry in Franco's Spain, II FUSCHROM Workshop on *Separation Strategies in Chromatography*, MWs as a fundamental tool in the analytical laboratory, the Earth at geologist's eye, contents as resource for encouraging interest to chemistry study, immobilized catalytic systems for asymmetric flow processes, three in one, three viruses and one micromurder, Zika, Chincungunya, and dengue, chemistry of the outer space, seeing the negative of the sky, the intergalactic medium and preparing students for practical sessions *via* laboratory simulation software.

The goal of this work is to initiate a debate by suggesting a number of questions on boom time for CE, creativity, whether it is all chemistry, and chemistry to improve people's lives, and providing, when possible, answers (A), hypotheses (H) and facts (F).

9.2 BOOM TIME FOR THE CIRCULAR ECONOMY AND CREATIVITY

A question and hypothesis were proposed on boom time for CE (*spiral*) and creativity:

Q1. The shape of the future is circular. However, how will that profit the waste industry?

H1. Creativity in science is the ability to get oneself into a fine mess.

9.3 IS IT ALL CHEMISTRY? CHEMISTRY TO IMPROVE PEOPLE'S LIFE

Puchades and Primo proposed questions/H/A on chemistry exclusiveness to improve life [8]:

Q1. Is it all chemistry?

Q2. What is the importance of the main elements that constitute matter?

Q3. What are the composition and nature of many products that people utilize daily?

Q4. How did chemistry improve people's lives in many aspects?

Q5. What are the products that people utilize?

Q6. How are they obtained?

Q7. What role do they play in people's daily lives?

Q8. How did they improve it?

Q9. How do chemical analysis techniques inform, protect, and allow improving one's quality of life?

H1. Chemophobia false myth: All that *smells* chemistry is bad [9].

H2. Positive meaning: *Between two persons there is chemistry.*

Q10. What is chemistry good for?

Q11. What does chemistry consist of?

Q12. What usefulness has chemistry?

Q13. What is matter?

Q14. What does matter consist of?
Q15. What usefulness has matter?
Q16. Does chemistry either help or damage people?
A16. Because there is an irrational phobia.
Q17. What does chemistry help people in?
Q18. What does it mean that chemistry serves to analyze?
Q19. What influence has chemical analysis on our daily life?
Q20. What importance has that analytical techniques could be developed?
Q21. Is there cocaine in all five-euro notes?
Q22. Is there cocaine in the environmental air?
Q23. Are there drugs in the rivers?
Q24. What does it mean to obtain chemical products?
Q25. Is it necessary?
Q26. What are the molecules?
Q27. How are the chemical compounds?
Q28. How is matter?
Q29. How are the molecules that constitute the products that people consume obtained?
Q30. What are they obtained from?
Q31. What raw materials are utilized?
Q32. What products do people consume?
Q33. What are the most consumed products?
Q34. Why do people depend so much on petroleum?
Q35. What is the petroleum?
Q36. How much petroleum do people consume?
Q37. What problems have the consumption of petroleum?
Q38. Are there some data on Spain?
Q39. How much natural gas (NG) and petroleum does Spain produce?
Q40. What does it happen with China?
Q41. What does it happen with the United States of America?
Q42. Besides fuel, what other things do people consume?
Q43. What do people obtain them from?
Q44. How are these chemical products that people consume, obtained?
Q45. Are they either natural or synthetic products?
Q46. What is chemical synthesis?
A46. To synthesize is to *build*.
Q47. Is vitamin-C (L-ascorbic acid) either natural or synthetic?
Q48. What is a detergent good for?

Q49. What properties must a molecule have to be detergent?
A49. It must have a hydrophilic head and a hydrophobic tail.
H3. Solubility old rule: *Like dissolves like.*
Q50. What is cyanogas $[Ca(CN)_2]$?
Q51. Have people alternatives to the use of pesticides?
Q52. What is a pheromone?
Q53. What types of insects pheromones are there?
Q54. How much pheromone does an insect produce?
Q55. How do people know it?
Q56. How can one know what molecule is the pheromone of an insect?
A56. By analytical techniques, especially by nuclear magnetic resonance (NMR)?
Q57. However, can people utilize pheromones against insect pests?
Q58. How is it used?
Q59. What concentration of pheromone is there in the air in a treatment of sexual confusion?
Q60. Is petroleum scarce?
A60. Petroleum to be burnt is *like eau-de-Cologne to light the barbecue.*
Q61. Why has the human body so much C and the Earth crust so little in relative composition?

9.4 SCIENTIFIC RESEARCH TO IMPROVE PEOPLE'S LIFE: AN INVESTMENT OF FUTURE

Corma raised the following questions on chemistry improving people's life [10]:

Q1. How will the more than 9,000 million inhabitants that will populate the Earth in 2050 eat?
Q2. How will people eradicate present diseases and those that they do not know yet?
Q3. How will people try to lessen some of the most serious consequences of climate change?
Q4. How to develop a sustainable chemistry sensitive to the entire complex problem?
Gabarrón reviewed Juan interviewing Corma and proposed Q/A/H/F on research improving people's life [11]:
Q5. Why not to go away again?

Q6. Why has chemistry so bad press?

A6. Because, sometimes, chemistry has not gone well and that has got consequences.

Q7. What examples would you give of bad experiences?

A7. When the preventive measures are not taken.

Q8. However, is not chemistry to blame?

A8. No, if the preventive measures were taken, there were not accidents.

H1. Research is to raise oneself questions and try to answer them.

Q9. Can everybody be a researcher?

A9. Up to a point, yes.

Q10. What is a *zeolite*?

A10. It is a *catalyst* with pores of diameter of molecules that one wants to react (molecular sieve): pore 1) controls molecule type that *goes in*; 2) selects how it *interacts* in active site (AS).

Q11. What is the size of the pore?

A11. It is 3.5–20Å: the size of the molecule that one wants to go in.

Q12. Does the catalyst sieve what molecule goes in and what molecule reacts?

A12. Different-diameters pores control molecular traffic: big/small molecules *via* wide/narrow pores meet at AS.

Q13. Why do you use a catalyst?

A13. To obtain desired product, avoid secondary/by-products/pollutants, save raw matter and improve selectivity.

F1. In *pharma* industry, one obtains 10 kg by-product/kg of drug, e.g., 12 kg→20 g by-product.

F2. In cars exhausts, there are catalysts, e.g., to put nanoparticles to eliminate NO_x in diesel motors.

Q14. Is there chance in research?

A14. Yes, that the company believe in research group, be in situation, understand basic research, etc.

Q15. Should basic research precede applied investigation?

A15. No, put all knowledge in a toolbox and, when there be a problem, attack it with the toolbox.

Q16. What is the role of companies?

A16. Companies must open the box of knowledge to solve their problems.

Q17. Can one give up fossil fuels today? Is it a chimera?

A17. One must optimize and, one day, people will spend oil only to make products, not to burn it.

Q18. What would you wish to do for the environment?

A18. Renewable energy (most important problems are sustainable water, food, health, and energy).

Q19. How do you look at spreading? Is it a duty of the scientific community?

A19. Yes of course! It must make it.

Q20. In addition, to write a book for ordinary mortals?

A20. The problem is time.

Q21. What is needed in the science and society relationship?

A21. More funding and politicians with more scientific training.

Q22. What about the use of CO_2?

A22. It is an ideal recyclable raw material for MeOH/carbonates/polycarbonates but one need H_2!

Q23. Are there novelties on H_2 by fusion energy?

A23. Fusion energy is unready.

Q24. What material are zeolites made from?

A24. They are silicoaluminates; let one think about clay with controlled crystallization to obtain certain-diameter pores; it is biodegradable.

F3. Pesticides sorbed on zeolites.

F4. Hormones that take part in love and sex are chemistry.

F5. Insect sexual confusion: pheromones sorbed on zeolites.

Q25. Banks have advisory boards on climate change. What would you advise them?

A25. Energy, concentration of energy and renewable energies.

Q26. Can a chemist be in agricultural chemistry?

Q27. Are farmers treating the environment badly?

A27. Excess of manure, insecticides, etc.

Q28. Do uncontrolled use of nitrates, insecticides, water, etc., improve with *ecological* agriculture?

A28. Masses agriculture is *intensive* rather than *ecological* but people can make insecticides that attack only one species.

Q29. What do you think about patents?

A29. *Research for what is needed here!* However, to repeat is not for researcher but to make new technology.

9.5 CHEMICAL SCIENCE FOR EVERYBODY

Pérez Castelló proposed questions and answer on chemical science for everybody [12]:

Q1. What image do people have of chemistry?
Q2. In kitchen, acetic acid, ammonium hydroxide, sodium chloride, etc., exist, do you know them?
Q3. Sociological Q (SQ)1. What is pH?
A3. *It is something that is added to cold creams.*
    She raised additional questions on the incombustible banknote:
Q4. Will a banknote burn in wet water?
Q5. Will a banknote burn in wet alcohol?
Q6. Will a banknote burn in wet a mixture of water and alcohol?
    She proposed the following additional question and answer:
Q7. What is the difference between physics and chemistry?
A7. Physics deals with the structure of matter; chemistry deals with the changes of matter.

9.6 CHEMISTRY AND ENERGY CHALLENGES OF THE 21ST CENTURY

Fontecave raised the following question on chemistry and energy challenges of the 21st century [13].

Q1. What energies for tomorrow?

    He proposed the following questions and answers on three problems:

Q2. Rise of energy demand in 35 years?
A2. From 16TW (2014) to 27TW (2050).
Q3. Exhaustion of the carbonaceous fossil reserves (oil, gas, coal)?
A3. Coal >200 years (?), gas <200 years, oil 50–100 years, gas of schist (?).
Q4. Nuclear?
A4. U (slow neutrons): 200 years; (rapid neutrons-Phoenix >5000 years.

    He raised the following additional questions and answer:

Q5. Temperature: reality and foresights?
Q6. Fossil energy?

Boom for Circular Economy and Creativity

Q7. Rejecting or storing?
Q8. Tomorrow, what transport?
Q9. H_2, future fuel?
Q10. CO_2, source of carbon?
A10. CO_2 provides CO, formic acid HCOOH, methanol CH_3OH (fuel?) and methane CH_4 (fuel?).
Q11. The scale of home consumption… how many liters of water per day?
Q12. What surface of photovoltaic (PV) panels?
Q13. The *hydrogen world*?

9.7 PRESENT AND FUTURE OF PLANT SAFE

Laborda raised the following questions on chemical ecology [14]:

Q1. How do insects communicate?
Q2. How can such communication be interfered?
Q3. However, what do you intend to eat?
Paniego proposed the following hypothesis on the present and future of plant safe [15]:
H1. Technicians are the *why* but farmers are the *how*.

9.8 QUANTIFYING AMORPHOUS CONTENTS IN MILLED POWDERS: CALORIMETRY

Gaisford proposed Q/A on quantifying amorphous contents in milled powders with calorimetry [16].

Q1. How do powder blends aerosolize?
Q2. Is the particle size of the sample the same every time and/or as the material used for calibration?
A2. If sample particle size is different from calibration, proportion of total heat change from adsorption varies.
Q3. Does the sample exhibit polymorphism?
A3. Calibration curve must be generated with material of the same polymorphic form being tested.
Q4. Is the material anomeric?
A4. Calibration curve must be generated with material of the same tested anomeric form (or ratio).

Q5. What plasticizer or solvent is being used?

A5. Plasticizer should be optimized for sample and the sample for prepared samples and calibration; the same solvent, for samples and calibration.

Q6. Is the method a comparison of wetting responses?

A6. Adsorption heat remains and the only way to correct is to dry/re-expose sample to plasticizer.

Q7. Can the sample form a hydrate or solvate?

A7. If sample forms a hydrate/solvate with plasticizer, then the power signal may become complex.

Q8. Could the sample have formed a hydrate or solvate before or after milling?

Q9. Were the samples used to make the calibration curve hydrates or solvates?

Q10. Is a hydrate or solvate formed directly on crystallization of the amorphous fraction?

Q11. Is the milled material sourced from different suppliers?

A11. Energy provided after brittle-ductile point will not cause size reduction but be dissipated by sample.

Q12. Why are milled materials usually partially amorphous?

Q13. Are samples analyzed at the same time after milling?

A13. Time between milling and amorphous-content measurement is critical (time↑, relaxation↑, amorphous content↓).

Q14. How was the amorphous reference material prepared?

A14. A calibration prepared with amorphous material is different from that with freeze-dried one.

Q15. Is percent amorphous content actually needed?

A15. To correlate excess energy with fine particle fraction (FPF) is better than percent amorphous content.

9.9 SAFETY STORAGE CABINETS AND SAFETY CONTROL OF FUME CUPBOARDS

Trallero raised questions and answer on the characteristics of safety storage cabinets [17]:

Q1. Are safety storage cabinets necessary?

Q2. Do they need extraction?

Q3. Do they need filters?

Q4. Only for inflammables?

Q5. In addition, in the case of acids and bases?

Q6. What is a safety storage cabinet for?

Q7. What are the differences between the different models of safety storage cabinets?

Q8. What does it mean REI 120?

A8. Fire resistance, tightness (étanchéité) and isolation during 120 minutes.

Q9. Safety storage cabinets for inflammable liquids, how do they function?

Q10. What does it do that a safety storage cabinet be safe?

Q11. What does a safety storage cabinet function?

Q12. Would not it be better to have a safety storage cabinet that allows one to evacuate the room and arriving the emergence teams?

Q13. Safety storage cabinets for bottles of gases, how do they function?

Q14. The safety storage cabinet, what is it for?

Q15. The safety storage cabinet; what differences are there between the different models?

Q16. The safety storage cabinet, what must one wonder before the purchase?

Q17. Does safety imply little flexibility?

Q18. The safety storage cabinet; what differences are there between the cabinets?

Q19. The safety storage cabinet; is cooling needed?

Q20. Resistance to fire?

Q21. What products is one going to store?

Q22. Must the safety storage cabinet have resistance to temperature?

Q23. Must one keep a low temperature?

Q24. Situation of the safety storage cabinet; is it possible to set up the cabinet where the user wants?

Q25. The safety storage cabinet for bottles of gases, what differences are there between the cabinets?

Q26. Are the rules observed?

Q27. Are you sure that your safety storage cabinet is in perfect conditions?

Q28. Do you think that your safety storage cabinet will allow you to avoid fire propagation and evacuate the laboratory?

Q29. Interior use?

Q30. How not to do it?

Q31. What is a dispensation system for?

174 Green Chemistry and Green Engineering

Q32. What appearance has the dispensation system?

Colás Marín raised questions on the control of the safety of fume cupboards [18]:

Q33. What is European Standard EN 14 175?
Q34. What is European Standard EN 14 175 useful for?
Q35. What must one search for in a fume cupboard?
Q36. Assays of type, what are they useful for?
Q37. *In situ* assays, why to verify the extraction flow can be an error?
Q38. How to avoid it?
Q39. How to install a fume cupboard?
Q40. Where to install a fume cupboard?

9.10 CATALYSTS LOOKING AFTER THE ENVIRONMENT

García Martínez proposed questions and answers on catalysts looking after the environment [19]:

Q1. How can science contribute to a more sustainable place?
Q2. Catalytic ammonia synthesis in homogeneous solution, biomimetic at last [20]?
Q3. How is H stored?
Q4. What are the problems of H (*the energy of the future*) in order to be used as a source of energy?
A4. Changing from fossil fuels to H has challenges: obtaining, purification, storage, and transport.
Q5. How is it gotten to manufacture 300Tm of catalyst?
A5. Making it economic, making it reliable and being sure of a great potential benefit.
Q6. CO_2 sequestration: burying CO_2 or can catalysis help?
A6. Problem is CO_2 concentration/mobility; catalysis rises value making, e.g., alcohols, and fuels.
A7. The problem is energy Carnot cycle: low energy CO_2 needs energy; a solution is *via* sunlight.
Q7. How do you see the problem of the international agreements to reduce CO_2 emissions?
A7. Every time more distrust exists, O_3-eating freons had alternatives, CO_2, not.

Boom for Circular Economy and Creativity

9.11 GRAPHENE (GR) SCIENCE AND TECHNOLOGY: ARE EXPECTATIONS COMING TRUE?

Menéndez proposed questions and answers on GR science and technology, and expectations [21]:

Q1. Graphene (GR), what do we know about it [22]?
Q2. Are expectations coming true?
Q3. Is GR, in fact, as powerful as it is said?
Q4. How does one obtain a material so prized?
Q5. Is it about a material addressed to high-technology applications or part of the day in day out?
Q6. How far has GR arrived in these years?
Q7. What ratio of what is investigated does it arrive to market?
A7. The bottleneck is the market.
Q8. How much can be foreseen?
A8. Companies research is secret and it will be seen only when it be launched to the market.
Q9. What ratio of what is investigated will substitute Si?
A9. The GR has applications neither in microelectronics but it will nor substitute Si.
Q10. When will GR be manufactured at industrial scale?
A10. It is already manufactured at industrial scale.
Q11. However, quality control?
A11. Problem is not production but use, volume of market; prize will be halved with more market.
Q12. It is the same case as CO_2 sequestration, where will it be stored?
Q13. What does GR or graphite contribute to composites?
A14. GR contributes greater strength (graphite is fragile); GR rises conductivity but graphite also.

9.12 FROM PETROLEUM TO SUNLIGHT: SOLAR-FUELS PHOTOCATALYTIC PRODUCTION

García Gómez proposed questions/answers from petroleum to sunlight and solar-fuels PCP [23]:

Q1. In the electric car, how electricity will be produced?
A1. With renewable sources.
Q2. In which material will electricity be stored?

A2. Energy is stored in: (1) V^{III}–V^V batteries but in industrial plants, not cars; (2) solar.

Q3. Will energy come from micro-organisms?

A3. H_2 from electrochemistry; but electrolizer and electrodes are expensive; my system is economic: in-gas $\rightarrow$ artificial-sun $\rightarrow$ out-gas.

Q4. Will energy come from CH_4 and cattle raising?

Q5. Interest?

A5. CO_2 can be treated in *CO_2 refineries* to produce now CH_4 and, in the future, CH_3OH.

Q6. H_2 is obtained from pure water, what happens from seawater?

A6. It goes worse, the material deactivates.

Q7. In addition to Au and Cu, have you tried other catalysts as Pd and Pt?

A7. Pd reacts with C to give Pd_2C; Pt goes worse than Au. The best are Au and Cu.

9.13 II FUSCHROM WORKSHOP ON SEPARATION STRATEGIES IN CHROMATOGRAPHY

In *Separation Strategies in Chromatography*, Berthod proposed Q/A on trends in countercurrent chromatography (CCC), liquid systems, equipments, and approaches [25].

Q1. Which is the most polar system [26]?

Q2. Why is a *liquid stationary phase* so interesting in separation methods?

A2. CCC can purify compounds on a *preparative* scale with *less solvent*.

Ramos proposed Q/A on GCxGC-TOF MS-hyphenation potential in environmental field [27].

Q3. What is *multidimensional chromatography* [multidimensional gas chromatography (MDGC)]?

A3. *Selective transfer* of a *pre-selected fraction from 1* chromatographic *separation medium to a 2^{ary} one*.

Q4. Is GCxGC necessary?

A4. Yes, for co-elutions and matrix effect.

Q5. Modulator?

Boom for Circular Economy and Creativity

A5. It puts fractions of the peak coming from the first-dimension (^{1}D) chromatography to ^{2}D one.

Q6. How many data points are necessary to characterize correctly a chromatographic peak?

9.14 MICROWAVES (MWS): A FUNDAMENTAL TOOL IN THE ANALYTICAL LABORATORY

GOMENSORO organized a Seminar on MWs as a basic tool in analytical laboratory. Palma raised Q on synthesis/calcination [28]:

Q1. What would it be ideal?

Q2. What is flexiWAVE?

Q3. What is synthWAVE?

Q4. Why to calcine *via* microwaves (MWs)?

Vélez raised the following questions on digestion *via* MWs in agro-food matrices [29]:

Q5. What?

Q6. In what?

Q7. With what?

Q8. Routine?

Q9. Research-development?

Q10. Service to third ones?

9.15 THE EARTH AT GEOLOGIST'S EYE

Ortega proposed the following questions and answers on the Earth at geologist's eye [30]:

Q1. How do geologists interpret landscape by reading in the written pages of rocks [31]?

A1. Convergence of forms between microstructures and macrostructures allows interpreting it.

Q2. Why are fossil mollusks found in the tops of the mountains [32]?

He proposed seven key questions (KQs) and answers on climate change:

Q3. KQ1. Is climate really changing?

A3. Yes.
Q4. KQ2. Is human activity contributing to climate change?
A4. Yes.
Q5. In what percentage?
Q6. KQ3. Is it necessary to slow down pollution?
A6. Yes.
Q7. KQ4. Is human activity giving rise to a heating unusual and unknown in Earth history?
A7. No.
Q8. KQ5. Are human activities causing climate change?
A8. No.
Q9. KQ6. Is human activity giving rise to accelerated changes much faster than geologic changes?
A9. No.
Q10. KQ7. Are scientists transmitting to public opinion the results of research on climate change?
A10. No.

9.16 CONTENTS AS RESOURCE FOR ENCOURAGING INTEREST TO CHEMISTRY STUDY

Chemistry does not enjoy good reputation in the social ambit. The expression *without chemistry* is used in a mountain of commercials as an advertising ploy for consumers. On the other hand, media emphasize the accidents that occur because of some undesirable reaction, while they devote little or nothing to emphasizing the applications that it presents in improving the quality of life that it shows in people's daily life, with applications in human health, food, environmental protection, discovery of useful materials, etc., all of which influences directly its *bad reputation*. In this, teaching staff in chemistry are responsible part on devoting most of the time to explain concepts and dedicating little to their applications, which, on the other hand, provide the opportunity that pupils discover a subject established in daily life, making it more attractive. In the habitual teaching of the subject, it is normal to stress products formulation and problem solving, mentioning little or nothing the use and applications, and even risks that the products and reactions present in daily life. Herradón proposed an H [33].

H1. If I were a pupil of baccalaureate, I would ask my center that taught me chemistry well.

He proposed the following questions on teaching chemistry well:

Q1. What is to teach chemistry well?
Q2. Much subject?
Q3. Not much subject and well learned?
Q4. Laboratory?
Q5. School hours?
Q6. Chemistry in *Physics and Chemistry*, in *Natural Sciences* or *alone*?

He proposed the following questions, answers, and hypotheses on what chemistry is:

Q7. What is chemistry?
A7. Science studying matter composition/structure/properties/changes, especially at atomic/molecular level.
Q8. How many chemistries are there?
A8. There is only one chemistry, a multidisciplinary and interdisciplinary science.
H2. Chemistry is related with two other sciences: biomedicine and materials science.
H3. Chemistry should be taught in a laboratory because it is an experimental science.

He proposed questions, answer, and hypotheses on chemistry as a part of culture:

Q9. Chemistry as a part of culture?
Q10. Why should people promote scientific culture?
A10. Pleasure to know/learn; criterion at expressing opinions; pseudo-/science distinction. Compare:
H4. Here we have the data, what conclusions can we take out from them?
H5. Here we have the conclusions, what data can we explain with them? He proposed the following questions and hypothesis on science, technology, and society:
Q11. Lack of scientific vocations, what is the cause?
Q12. Is the utility of science perceived?
Q13. Boring subject?
Q14. Difficult?
Q15. Quality and quantity of studies in secondary education/baccalaureate?
Q16. Social recognition of scientists?

Q17. When does scientific vocation begin?

Q18. When should one begin studying science?

H6. (Echenique). Science is the greatest collective work in the history of humanity.

Q19. What image has an adolescent of scientists?

He proposed the following questions on chemical education:

Q20. When to begin?

Q21. In primary education?

Q22. What can be taught in the laboratory in secondary education?

He proposed questions, hypotheses, and answer on what chemistry is and what it is for:

Q23. What is chemistry?

Q24. What is chemistry for?

H7. (Pauling). If you want to study function, study structure.

H8. (Atkins). The great ideas in chemistry [34]: matter consists of *ca.* 100 elements; elements are composed of atoms; the orbital structure of atoms accounts for their periodicity; chemical bonds form when electrons pair; shape is central to function; molecules attract and repel each other; energy is blind to its mode of storage; reactions fall into a small number of types; reaction rates are summarized by rate laws.

Q25. What is chemistry?

H9. Chemistry creates its own object (*synthetic chemistry*).

H10. Natural *vs.* artificial dilemma.

H11. (United Nations). The Global Goals for Sustainable Development (*cf.* Figure 9.1): no poverty; zero hunger; good health and well-being; quality education; gender equality; clean water and sanitation; affordable and clean energy; decent work and economic growth; industry, innovation, and infrastructure; reduced inequalities; sustainable cities and communities; response consumption and production; climate action; life below water; life on land; peace and justice strong institutions; partnerships for the goals.

Q26. What is to be cultivated?

H12. The *two* [35] and *five cultures*: Humanistic/Scientific (Literature/Art)/Social sciences (economy, sociology, psychology)/(Pure sciences/Applied sciences).

Q27. Why should people promote scientific culture?

A27. In order to help to increase culture.

Q28. Nanoscience, the science of teeny tiny things?

Q29. How did Leonardo da Vinci paint those faces so perfect?
H13. I do not know how the future will be.
H14. I do not know either how the future of science will be.
H15. Without science, there is no future.
H16. (Descartes). I shall swap all that I know for half of what I ignore.

FIGURE 9.1 United Nations: The global goals for sustainable development.

9.17 IMMOBILIZED CATALYTIC SYSTEMS FOR ASYMMETRIC FLOW PROCESSES

Pericàs Brondo proposed questions/H/A on immobilized catalytic systems (*cf.* Figure 9.2) [36]:

Q1. Why do flow processes present optimal sustainability characteristics?
H1. Reduce, reuse, and recycle (3R).
H2. From batch (*reaction time*) to continuous flow (*residence time*) processes.
Q2. From batch to continuous flow processes with polymer-supported catalysts, what is required?
A2. High catalytic activity; high chemical and mechanical stability.

Q3. Are catalytic enantioselective flow processes difficult to optimize?
Q4. What is the expected reward?

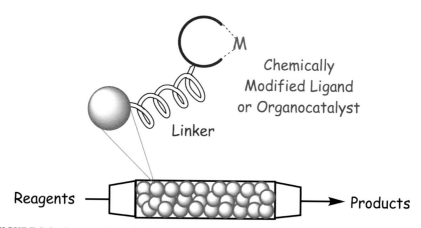

FIGURE 9.2 Supported catalysis: An approach to suitable synthetic processes.

H3. The importance of mixing.
Q5. How to avoid Cu leaching in immobilized tris(triazolyl)methane (TTM)-Cu complexes?
Q6. Why supporting chiral phosphoric acids?

9.18 3 IN 1 (3 VIRUSES AND 1 MICROMURDER): ZIKA, CHIKUNGUNYA, AND DENGUE

Mateo raised Q on three in one, 3 viruses and 1 micromurder, Zika, Chikungunya, and dengue [37].

Q1. What can people do before this tsunami that continues growing?
Q2. Can chemistry solve it?

9.19 CHEMISTRY OF THE OUTER SPACE

Figueruelo proposed questions and answers on the chemistry of outer space (*cf.* Figure 9.3) [38]:

Q1. Is their presence of prebiotic molecules in the interstellar space (IS)?

Q2. Is there extraterrestrial life?
Q3. Radiation that one detects from molecule X delayed hundreds of years, does it continue there?

FIGURE 9.3 Interstellar space: Horsehead Nebula (Barnard 33, B33).

Q4. If people jointly detect X and precursor Y, can they study reaction $X \rightarrow Y$, kinetics, past, and future?
Q5. If past events are similar to future ones, can one extrapolate to their future?
Q6. Are H compounds the unique molecules in the Universe?
A6. No, they are not.
Q7. In particular, or is general the chemistry on the Earth?
Q8. Is chemistry the derivative of life?
Q9. Is it the cause of life?
Q10. The probability of life in other parts of the Universe, what have chemists to say?
A10. Chemistry is without big differences.
Q11. Are there prebiotic molecules in IS?
Q12. Sugars?

Q13. Amino acids?

Q14. Cyclic?

A14. Yes, but they do not seem precursors but pyrimidine (in prebiotic chemistry laboratory).

Q15. In addition, what can people do in this field?

A15. (1) Search/detection of new molecules (better detection methods, e.g., Atacama Large Millimeter Array (ALMA); (2) better space telescopes, e.g., James Webb Space Telescope (JWST); (3) intensify in laboratory better instruments with ultrahigh vacuum and variable temperature; (4) hyphenated mass spectrometry-Fourier-transform infrared spectroscopy (MS-FTIR).

9.20 SEEING THE NEGATIVE OF THE SKY: THE INTERGALACTIC MEDIUM

Fernández Soto raised questions on seeing the negative of the sky and the intergalactic medium [39]:

Q1. What are the diffuse clouds of gas in the intergalactic medium?

Q2. How do the diffuse clouds of a gas behave in the intergalactic medium?

Q3. How are the diffuse clouds of gas that emit little or no light in the intergalactic medium studied?

9.21 PREPARING STUDENTS FOR LABORATORY VIA SIMULATION SOFTWARE

A series of laboratory-themed simulations developed by Learning Science Ltd. were integrated into a first-year laboratory module at the University of Leicester, which simulations allow students to attempt the experiments they will do in the laboratory in a risk-free way that provides the opportunity to make mistakes and learn how to correct them *via* the generated immediate feedback [40]. High student engagement was observed during their pilot, and student end-of-module comments were positive for the technological enhancement.

Boom for Circular Economy and Creativity   185

9.22 FINAL REMARKS

From the present results and discussion, the following final remarks can be drawn:

1.   Creativity in science is the ability to get oneself into a fine mess.
2.   Chemophobia false myth: All that *smells* chemistry is bad.
3.   Positive meaning: *Between two persons there is chemistry*.
4.   Solubility old rule: *Like dissolves like*.
5.   Research is to raise oneself questions and try to answer them.
6.   Technicians are the *why* but farmers are the *how*.
7.   New (social, etc.), ethos: Spain–US Agreement (Pact of Madrid, 1953).
8.   Better things for better living through chemistry.
9.   Not all fascist science is bad but excellent science requires social freedom (imagination, etc.).
10. Practically, no historical rhetoric exists and, in Spanish-Civil-War Generation, totals rupture.
11. Ingenuity and particular nature of chemistry for armor-plating itself (compare it to Darwinism).
12. Tacit knowledge: intangible.
13. Operation Paperclip: many German scientists recruited in post-Nazi Germany and taken to the US for government employment.
14. There are historical processes that require a long time.
15. If I were a pupil of baccalaureate, I would ask my center that taught me chemistry well.
16. Chemistry is related with two other sciences: biomedicine and materials science.
17. Chemistry should be taught in a laboratory because it is an experimental science.
18. Here we have the data, what conclusions can we take out from them?
19. Here we have the conclusions, what data can we explain with them?
20. Science is the greatest collective work in the history of humanity.
21. If you want to study function, study structure.
22. The great ideas in chemistry: matter consists of 100 elements; elements are composed of atoms; the orbital structure of atoms accounts for their periodicity; chemical bonds form when electrons pair; shape is central to function; molecules attract and repel each other; energy is blind to its mode of storage; reactions fall into a small number of types; reaction rates are summarized by rate laws.

23. Chemistry creates its own object (*synthetic chemistry*).
24. Natural *vs.* artificial dilemma.
25. The Global Goals for Sustainable Development: no poverty; zero hunger; good health and well-being; quality education; gender equality; clean water and sanitation; affordable and clean energy; decent work and economic growth; industry, innovation, and infrastructure; reduced inequalities; sustainable cities and communities; response consumption and production; climate action; life below water; life on land; peace and justice strong institutions; partnerships for the goals.
26. The *two* and *five cultures*: Humanistic/Scientific (Literature/Art)/ Social sciences (economy, sociology, psychology)/(Pure sciences/ Applied sciences).
27. I do not know how the future will be.
28. I do not know either how the future of science will be.
29. Without science there is no future.
30. I shall swap all that I know for half of what I ignore.
31. Reduce, reuse, and recycle.
32. From batch (*reaction time*) to continuous flow (*residence time*) processes.
33. The importance of mixing.

ACKNOWLEDGMENTS

The authors thank support from Fundacion Universidad Catolica de Valencia San Vicente Martir (Project No. 2019-217-001UCV).

KEYWORDS

- **chemical science**
- **countercurrent chromatography**
- **energy challenges**
- **mass spectrometry**
- **milled powders**
- **safety storage cabinet**
- **scientific research**

REFERENCES

1. Torrens, F., & Castellano, G., (2018). QSPR prediction of retention times of methylxanthines and cotinine by bioplastic evolution. *Int. J. Quant. Struct.-Prop. Relat., 3*, 74–87.
2. Torrens, F., & Castellano, G., (2018). Molecular classification of caffeine, its metabolites, and nicotine metabolite. In: Ul-Haq, Z., & Wilson, A. K., (eds.), *Frontiers in Computational Chemistry* (Vol. 4, pp. 184–224.). Bentham: Hilversum, Holland.
3. Torrens, F., & Castellano, G., (2017). QSRP prediction of retention times of chlorogenic acids in coffee by bioplastic evolution. In: Kandemirli, F., (ed.), *Quantitative Structure-Activity Relationship* (pp. 45–61). InTechOpen: Vienna, Austria.
4. Torrens, F., & Castellano, G., (2020). QSPR prediction of chromatographic retention times of tea compounds by bioplastic evolution. In: Justino, G., (ed.), *Tea – Chemistry and Pharmacology. InTechOpen*: Vienna, Austria; pp. 161–173.
5. Torrens, F., & Castellano, G., (2020). Elemental classification of tea leaves infusions: Principal component, cluster, and meta-analyses. In: Justino, J., (ed.), *Tea: From Chemistry to Pharmacology. In: Justino, G., (ed.), Tea – Chemistry and Pharmacology. InTechOpen:* Vienna, Austria; pp. 292–307.
6. Torrens, F., & Castellano, G., (2019). Beer, all a science, alcohol: Analysis and quality testing at multiple brewery stages. In: Vakhrushev, A. V., Mukbaniani, O. V., & Susanto, H., (eds.), *Chemical Technology and Informatics in Chemistry with Applications* (pp. 99–108). Apple Academic-CRC: Waretown, NJ.
7. Torrens, F., & Castellano, G., Beer molecules, sensory, bioproperties and street binge drinking. In: Haghi, A. K., (ed.), *Practical Applications of Physical Chemistry in Food Science and Technology*. Apple Academic-CRC: Waretown, NJ, in press.
8. Puchades, R., & Primo, J., (2019). *Personal Communication*.
9. Bensuade-Vincent, B., & Simon, J., (2012). *Chemistry: The Impure Science*. Imperial College Press: London, UK.
10. Corma, A., (2019). *Personal Communication*.
11. Gabarrón, M., (2016). La ciència, una inversió de futur. Science, an investment in the future. *Métode, 2016*(90), 1–1.
12. Pérez Castelló, E., (2019). *Personal Communication*.
13. Fontecave, M., (2019). *Personal Communication*.
14. Laborda, R., (2019). *Personal Communication*.
15. Paniego, C., (2019). *Personal Communication*.
16. Gaisford, S., (2017). Considerations when quantifying amorphous contents in milled powders with calorimetry. *Pharm. Technol. Eur., 29*(4), 21–24.
17. Trallero, P., (2019). *Personal Communication*.
18. Colás-Marín, F., (2019). *Personal Communication*.
19. García-Martínez, J., (2019). *Personal Communication*.
20. Broda, H., & Tuczek, F., (2014). Catalytic ammonia synthesis in homogeneous solution: Biomimetic at last. *Angew. Chem., Int. Ed., 53*, 632–634.
21. Menéndez, R., (2019). *Personal Communication*.
22. Menéndez, R., & Blanco, C., (2004). *Graphene*; ¿QuéSabemos de? No. 57, CSIC-Catarata: Madrid, Spain.
23. García Gómez, H., (2019). *Personal Communication*.
24. Nieto-Galan, A. *Chemistry in Franco's Spain*. Work in preparation.

25. Berthod, A., (2017). *Book of Abstracts, Second Workshop on Separation Strategies in Chromatography* (p. 1). Burjassot, València, Spain. Universitat de València: València, Spain.

26. Berthod, A., (2002). *Countercurrent Chromatography: The Support-Free Liquid Stationary Phase.* Comprehensive analytical chemistry no. 38, Elsevier: Amsterdam, The Netherlands.

27. Ramos, L., (2017). *Book of Abstracts, Second Workshop on Separation Strategies in Chromatography* (pp. 1–2). Burjassot, València, Spain. Universitat de València: València, Spain.

28. Palma, Á., (2017). *Book of Abstracts* (pp. 1–4)*. Microwave: A Fundamental Tool in the Analytical Laboratory.* València, Spain, GOMENSORO: València, Spain.

29. Vélez, D., (2017). *Book of Abstracts* (pp. 1–7)*. Microwave: A Fundamental Tool in the Analytical Laboratory.* València, Spain, GOMENSORO: València, Spain.

30. Ortega, E., (2019). *Personal Communication.*

31. Ortega, G. E., (2015). *Silver Tears: Multi-temporal Story-Fiction Story.* Chiado: Madrid, Spain.

32. Gould, S. J., (1998). *Leonardo's Mountain of Clams and the Diet of Worms.* Harmony: Danvers, MA.

33. Herradón, B., (2019). *Personal Communication.*

34. Atkins, P., (1999). Chemistry: The great ideas. *Pure Appl. Chem., 71*, 927–929.

35. Snow, C. P., (1959). *The Two Cultures and the Scientific Revolution.* Cambridge University Press: New York, NY.

36. Pericàs, B. M., À. *Personal Communication.*

37. Mateo, P., (2019). *Personal Communication.*

38. Figueruelo, J. E., (2019). *Personal Communication.*

39. Fernández, S. A., (2019). *Personal Communication.*

40. Blackburn, R. A. R., Villa-Marcos, B., & Williams, D. P., (2019). Preparing students for practical sessions using laboratory simulation software. *J. Chem. Educ., 96*, 153–158.

CHAPTER 10

Application of Engineered Nanomaterials in Environmental Protection and the Visionary Future

SUKANCHAN PALIT

Assistant Professor (Selection Grade), Department of Chemical Engineering, University of Petroleum and Energy Studies, Energy Acres, Post-Office-Bidholi via Premnagar, Dehradun – 248007, Uttarakhand, India; 43, Judges Bagan, Haridevpur (PO), Kolkata – 700082, India, E-mails: sukanchan68@gmail.com, sukanchan92@gmail.com, sukanchanp@rediffmail.com

ABSTRACT

The world of science and technology is moving towards a visionary era of scientific introspection and scientific vision. The challenges and the vision of environmental engineering science are immense and versatile. This treatise deeply comprehends with insight and deep scientific understanding of the truth of environmental remediation and unveils the areas of conventional and non-conventional environmental techniques. The authors in this treatise review the success of environmental engineering science in degrading recalcitrant organic and inorganic compounds in industrial wastewater and drinking water. Heavy metal and arsenic groundwater contamination are a veritable scientific burden in many developing and developed nations around the world. Mankind's immense scientific and knowledge prowess, the provision of clean drinking water, and the scientific excellence are all the torchbearers towards a new scientific order in environmental engineering science. Today nano-science and nanotechnology are integrated with diverse areas of science and engineering which includes environmental engineering and water purification. The authors in this chapter elucidate the various nanotechnology techniques in water/

wastewater treatment. Heavy metal and arsenic groundwater contamination in developing and developed nations around the world are veritably destroying the scientific firmament of might and vision. Technology has few answers to these ever-growing concerns. In this chapter, the author also delineates the different environmental engineering techniques in tackling arsenic groundwater contamination. The other areas of research pursuit in this treatise are the application of nanotechnology and nanomaterials (NMs) in environmental protection. Scientific vision and scientific divination are evolving at a rapid pace in this century. These areas of modern science and modern engineering are elaborated in deep detail in this chapter.

10.1 INTRODUCTION

Environmental engineering today is in the middle of deep scientific vision and comprehension. Industrialization and the march of human civilization have ushered in a disaster of immense proportion. The global climate change and the lack of clean drinking water have veritably challenged the scientific fabric. The utmost necessity of the hour is to target the areas of environmental engineering which are yet to be unexplored. Nanotechnology and nanomaterials (NMs) applications in environmental remediation need to be investigated with vast scientific vision and deep scientific introspection. Millions of people around the world are without clean drinking water. Lack of pure drinking water and heavy metal and arsenic contamination of groundwater are matters of immense and visionary concerns today. This is one of the pervasive problems afflicting people throughout the world. 1.2 billion people lack access to safe drinking water, 2.6 billion have little or no sanitation, and millions of people die annually (3900 children a day) from diseases transmitted through unsafe water or human excreta. Here comes the need of a deep scientific contemplation and scientific profundity in the field of drinking water and groundwater treatment. Mankind's immense scientific and knowledge prowess, the success of environmental engineering science and the needs of environmental sustainability will surely open up new doors of innovation and instinct in decades to come. In this similar vein, technological and scientific verve, validation, and vast motivation are the pillars of research pursuit in environmental remediation today. In this chapter, the authors reiterate the vision of environmental protection science, new methodologies, and new innovations. Global climate change needs to be uprooted with immediate effect. This scientific vision is delineated with insight in this chapter. The areas of sustainability such as environmental,

Application of Engineered Nanomaterials in Environmental Protection 191

energy, social, and economic are dealt with vision, might, and scientific determination in this chapter. Environmental sustainability or green sustainability is the absolute need of scientific progress today. The truth of science, the ingenuity of environmental remediation, and the scientific humanism are all the forerunners towards a new visionary era in environmental engineering [20–26].

10.2 THE AIM AND OBJECTIVE OF THIS STUDY

Environmental engineering and chemical process engineering in today's scientific world are in the avenues of new restructuring. Global water problems are in the process of severe devastation. Thus the need of new innovations, new research forays, and newer scientific profundity. The importance of NMs and engineered NMs thus comes to the forefront of any research endeavor today. Thus in a similar vein, environmental protection science is aligned with nanotechnology and NMs. The author deeply discussed the recent developments in NMs, nanotechnology, and their applications in environmental remediation. Application of engineered NMs in environmental protection is the utmost need of the hour. This chapter will widely envision the success and profundity of the science and engineering of engineered NMs. Integrated water resource management and urban water management should be integrated with innovations in nanotechnology today. The world of the science of engineered NMs will surely open up new doors of innovation in the field of application and scientific validation of environmental engineering techniques. The main vision of the chapter is to envision the field of engineered NMs in its applications in diverse areas of science and technology. Integrated water resource management and natural resource management are in a situation of immense grave consequences globally today. Thus, technological vision and a vast scientific validation is the utmost need of the hour. Water and wastewater treatment tools are also the need of the moment. This is also another hallmark of this well-ventured treatise [25, 26].

10.2.1 WHAT DO YOU MEAN BY ENGINEERED NANOMATERIALS (NMS)?

NMs and engineered NMs are today the visionary areas of research pursuit. Engineered NMs are chemical substances or materials that are engineered with particle sizes between 1 to 100 nanometers in at least one dimension. Mankind's vast scientific vision and deep scientific redeeming and the necessity of environmental protection will surely today open up new corridors of invention and innovation in the field of both nanotechnology and environmental protection. It is highly established that engineered NMs derive many functional advantages from their vast and unique physical and chemical properties. These novel properties have resulted in a high interest in discoveries and innovations across many industrial, commercial, and medical sectors. The world of scientific challenges and the need of scientific validation will surely broaden the areas of interdisciplinary areas of science and engineering such as integrated water resource management and water quality management. This is the pivot of this chapter. Many of the properties such as particle size, surface area, and surface reactivity also influence their inherent hazard and highly threaten the health of workers, communities, and the environment. Chemical process safety and chemical process design engineering are the needs of research endeavors in NMs and nanotechnology today. In this chapter, the author deeply and comprehensively elucidates on the needs of engineering science and technology in the successive build-up of nanotechnology and nanoengineering. Humanity and scientific progress are in a difficult situation today as environmental disasters and global climate change are challenging and denigrating the scientific vision today. This chapter forcefully relates the success of the science of engineered NMs in the true emancipation of science and engineering [25, 26].

NMs are materials which have the structural components sized from 1 to 100 nm. They have unique properties when compared with other conventional materials, such as mechanical, electrical, optical, and magnetic properties due to their small size and higher specific surface area (SSA). NMs describe, in principle, materials of which a single unit is sized between 1 to 1000 nanometers but usually is 1 to 100 nm. NMs research takes a material science and composite science-based approach to nanotechnology, envisioning advances in materials metrology and vast synthesis which have been developed in deep support of microfabrication research. Materials with structure at the nanoscale (NS) level often have unique optical, electronic, or mechanical properties. This is the major scientific vision behind application of NMs and engineered NMs. NMs are slowly commercialized and slowly beginning to emerge as industrial and scientific commodities. Scientific imagination, deep scientific transcendence and the vast scientific doctrine of nanotechnology are thus opening corridors of innovation and engineering instinct in decades to come. A new door of innovation and a new window of scientific and engineering emancipation will thus usher in a new age in nanotechnology research. In this chapter, the author deeply delves into the scientific success and the engineering profundity in the field of NMs research [25, 26].

10.3 ENVIRONMENTAL PROTECTION, ENVIRONMENTAL SUSTAINABILITY, AND THE VISION FOR THE FUTURE

Sustainable development and sustainability engineering are the pillars of scientific endeavor globally today. A well-researched treatise is the need of the hour. The scientific vision and the success of science of sustainability will be the cornerstone of all research endeavors today. Energy and environmental sustainability are the vision of science and engineering today. Global climate change and global warming are the scientific burdens of civilization. Depletion of fossil fuel resources is devastating the scientific and engineering landscape. Thus the need of a concerted effort from civil society, scientists, engineers, and governments around the world. Environmental and green sustainability will surely usher in a new era in the field of environmental engineering science. In this chapter, the author, with scientific vision and conscience targets the areas of nanotechnology and environmental protection science which needs to be investigated at the utmost. The true vision and abundance of science and engineering thus will be re-envisioned with the progress of civilization. Throughout the world, a deep scientific vision and scientific forbearance are required in the scientific progress of environmental engineering. The sagacity of environmental engineering science and chemical process design engineering needs to be re-envisioned as civilization treads forward. This chapter deeply elucidates the success of environmental protection science in the greater emancipation of environmental sustainability. In this age of modern science, sustainable development in the field of energy and environment is the scientific imperative of today. There are no other answers beyond environmental and green sustainability. The true vision of Dr. Gro Harlem Brundtland, the former Prime Minister of Norway on the science of sustainability needs to be reemphasized and revamped as mankind treads forward. This is the era of nuclear technology and space science. A brief description on the challenges of NMs in environmental protection is the cornerstone of this chapter [21–26].

10.4 MODERN SCIENCE AND THE MARCH OF ENVIRONMENTAL ENGINEERING

Modern science today is in the middle of deep scientific comprehension and immense scientific honing. Rapid and vigorous industrialization, the rupture of environment, and the depletion of fossil fuel resources are changing the face of science and engineering today. Modern science and modern human society are thus in the crucial juxtaposition of comprehension and deep introspection. Today interdisciplinary areas of research and engineering are the need of scientific progress and deep academic rigor today. Science and technology of environmental protection and water purification are also in the middle of vast, varied, and sound research and development initiatives. Modern science and arsenic and heavy metal decontamination of groundwater and surface water are the scientific imperatives of greater vision, might, and profundity. Bangladesh and the West Bengal state of India are grappling this situation with the help of western countries. Mankind's immense scientific truth and the emancipation of engineering science will all be the true forerunners towards a newer visionary era in environmental protection. The need of environmental and green sustainability and the advancements of green nanomaterial are in the process of new scientific regeneration.

10.5 THE SCIENTIFIC DOCTRINE OF NANOMATERIALS (NMS) AND ENGINEERED NANOMATERIALS (NMS)

NMs and engineered NMs are today applied in every areas of environmental engineering today. The need of civilization and the scientific alarm of nanotechnology are the wide vision of health aspects of NMs. Environmental remediation and the science of nanotechnology are today undivided areas of science and engineering. The deep scientific doctrine, the need of scientific imagination, and the vision to move forward are the torchbearers towards a new age in nanotechnology and engineered NMs. Today industrialization and the march of human civilization have veritably destroyed the environment. Technological advancements and scientific vision are today in the middle of vast disaster and immense scientific comprehension. Nanotechnology today is in the forefront of scientific emancipation. This chapter will surely unveil the thoughts of science and engineering in the field of engineered NMs. The area of urban water management and sustainable water resource management are in the middle of immense crisis. Nanotechnology and nano-engineering

Application of Engineered Nanomaterials in Environmental Protection 195

will surely the problems of this proportion. Civilization's scientific stance, the ardor and candor of scientific validation and the needs of green sustainability will surely open new avenues of emancipation in integrated water resource management.

10.6 SCIENTIFIC SAGACITY AND THE APPLICATION OF NANOTECHNOLOGY

The deep scientific sagacity of nanotechnology and the vast engineering profundity of nano-engineering are changing the face of scientific endeavor today. Water treatment, industrial wastewater treatment and the vast domain of water purification are today changing the vast scientific firmament. In this chapter, the author rigorously points towards the ingenuity in the field of NMs. Civilization is today in a disastrous state as environmental remediation stands as a pivotal issue. The vicious issue of climate change and global warming will surely open newer thoughts and new future recommendations in the field of environmental remediation. Research pursuit in the field of nanotechnology is today surpassing vast and versatile scientific frontiers. In this research endeavor, the author deeply elucidates the scientific barriers, the scientific fortitude, and the scientific redeeming in the field of applications of NMs and engineered NMs in environmental engineering science. The civilized society today is in the middle of deep scientific transformation and deep scientific introspection as regards application of nanotechnology. Water purification science and water technology are the torchbearers towards a new field of environmental engineering. Scientific transcendence and deep scientific intuitiveness will surely unravel the intricacies and the scientific and engineering barriers in the field of environmental remediation [21–23].

10.7 THE APPLICATION OF NANOMATERIALS (NMS) IN ENVIRONMENTAL PROTECTION

Technology and engineering science in the present-day human civilization needs to be reorganized and revamped as regards environmental remediation. NMs and engineered NMs are the scientific imperatives of modern science and modern civilization. The application areas of NMs in diverse areas of science and engineering are immense and pivotal. Civilization's immense knowledge prowess, the success of scientific truth, and the needs of green or environmental sustainability will all be the scientific truth and scientific vision of tomorrow. The redeeming of science and the future recommendations in the field of environmental protection will surely today usher in a newer age in both chemical process design engineering and environmental engineering. Both integrated water resource management and urban water quality management are in the midst of immense scientific introspection and deep contemplation. The scientific sagacity in the applications of NMs and its health effects are of immense comprehension in the path of academic and scientific rigor today. Engineering and technological validation will veritably open up new branches in the field of water resource management and water quality management [24–26].

10.8 THE SUCCESS OF SCIENCE OF ENVIRONMENTAL SUSTAINABILITY

Environmental sustainability and environmental engineering are today undivided domains of science and engineering. The visionary words of Dr Gro Harlem Brundtland, former Prime Minister of Norway on the science of sustainability needs to be reorganized and revamped as civilization treads forward. Nanotechnology and engineered NMs are the wonders of science and technology today. The success and ingenuity of science and engineering of environmental sustainability lies in the hands of engineers, scientists, governments, and the civil society. Scientific honing and engineering redeeming are the needs of research pursuit today. Millions of people around the world lack provision of clean drinking water. Civilization's immense scientific stance, the world of scientific and engineering validation and the needs of water for humanity will surely open up new avenues in the field of environmental engineering and environmental remediation. Nanotechnology and NMs are the minarets of scientific endeavor today. The need is of a concerted effort from the civil society. Today environmental sustainability is integrated with every branch of environmental engineering science such as integrated water resource management and water quality management. The success of the science of sustainability is vast and versatile as mankind moves forward. The need of water quality management and water resource management are immense as science and civilization moves forward. This chapter opens up new thoughts and newer future recommendations in the field of environmental protection.

10.9 RECENT SCIENTIFIC RESEARCH PURSUIT IN THE FIELD OF NANOTECHNOLOGY AND ENVIRONMENTAL REMEDIATION

Research pursuit and scientific emancipation and redeeming are the needs of human progress globally today. Nanotechnology and environmental remediation today should be integrated with each other if fruits of science are to be achieved. In this section, the author rigorously points towards the recent scientific pursuit in the field of nanotechnology and nano-engineering applications in environmental protection. The verve of science and the world of scientific validation will surely open up a new era in the field of engineered NMs. Health issues and public health engineering problems are the pivots of the use of NMs in human society today. The march of the science of nanotechnology and nano-engineering should be integrated with the vast world of health science. The author in this section widely opens a new era in the field of nanotechnology.

Inter-American Development Bank Report [1] deeply elucidated with vast scientific insight the future of water science and technology. This report is a collection of essays on disruptive technologies that may transform the water sector in the next 10 years [1]. A deep scientific vision and ingenuity are the hallmarks of this report. Resource recovery and water technology are the needs of human civilization and human progress today. This report forcefully delineates disruptive innovations in the water sector, positive water technology disruptions by 2030, and the digital future of water science and engineering [1]. Civilization's immense scientific steadfastness, the ingenuity, and the foresight of environmental protection and water technology will surely open up new doors of intuitiveness and insight in science and technology in years to come [1]. Innovation and scientific and engineering intuitiveness are quickly and inevitably changing the vision of infrastructure services [1]. Waterfalls under essential infrastructure services. Processes and challenges are being transformed and frontiers are shifted. This volume of research expertise compiles the answers to this vital question from four experts in water science and environmental remediation. Attention and deep research introspection are also concentrated on regulation, policy, and the market strategy in the adoption and mainstreaming of water science and technology [1]. The authors deeply stressed on water and wastewater systems and the decentralized networks that rely on remote sensing and digital technologies to control water quality and quality parameters to ensure safe and affordable drinking water. The challenges of water engineering and technology are vast and versatile in today's human civilization [1]. The over-arching goal of this report elucidates on the success, the contemplation, and the vision of water technologies till 2030. The world of challenges in integrated water resource management and urban water quality management are of immense importance today. Desalination and water treatment are the challenges of developing and developed nations around the world. Water stressed countries around the globe are in the war footing in enhancing green sustainability and sustainable water practices [1]. Climate change and water scarcity are changing the face of human civilization today. The situation of water scarcity is grave and truly alarming in developing nations such as India and Bangladesh. Arsenic and heavy metal groundwater poisoning are challenging the entire South Asian countries such as Bangladesh and the state of West Bengal, India [1]. The barriers and intricacies of sustainable development are vast and varied. This report successfully portrays the needs and scientific imperatives of global research prowess in water science and technology. The challenges and the targets of urban water quality management and the vast vision behind it are elucidated in detail in this report [1].

Kunduru et al. [2] with deep scientific insight elucidated nanotechnology for water purification and the application of nanotechnology methods in water purification. Water is veritably an important asset for humanity. Potable water supply is a need of human race. However, civilization is far from meeting global demands. This issue will enhance in future [2]. Here comes the importance of nanotechnology and its innovations. Demand increases due to population growth, climate change, and deteriorating water quality. Integrated water resource management and water quality management will be the scientific success of this century. Only 2.5% of the world's oceans and seas harness freshwater (freshwater having salts concentration less than 1 g/l [2]. However, 70% of freshwater is frozen as eternal ice [2]. Only less than 1% of freshwater is useful for drinking. Therefore, water treatment and wastewater treatment technologies should be implemented in these areas [2]. The problem is severe in developing nations and sub-Saharan countries around the world. Today the global situation is extremely devastating and necessity grows as innovation treads forward. Today is also a technology-driven society. The concepts of reuse, recycle, and repurpose are the imminent needs of the day. Water contaminants may be organic, inorganic, and biological in nature. Some contaminants are toxic and carcinogenic and have disastrous effects on human health and eco-systems [2]. Civilization and science need to be re-envisioned as environmental protection science treads forward. Some heavy metals such as arsenic are disastrous water pollutants. South Asia such as Bangladesh and the West Bengal state of India are in

Application of Engineered Nanomaterials in Environmental Protection 199

the grapple of a disastrous environmental issue that is arsenic groundwater poisoning [2]. Here comes the need of innovation and scientific intuitiveness. Other heavy metals which are dangerous are cadmium, chromium, mercury, lead, zinc, nickel, copper, and so on; they have serious toxicities. Thus, human scientific transcendence and scientific redeeming are in a crucial juncture. Nanotechnology is the scientific imperative of today's science and engineering. The authors in this chapter described in details importance of nanotechnology in water purification, major limitations, applications of nanotechnology in water or wastewater treatment, overview of different NMs in water and wastewater treatment, nanoadsorbents, carbon-based nanoadsorbents, and membranes and membrane processes [2]. The other areas of immense scientific introspection are advanced oxidation processes such as photocatalysis and the areas of antimicrobial NMs in disinfection and microbial control [2]. Nanoantimicrobial polymers and its application areas are the other hallmarks of this chapter. Scientific vision, vast scientific steadfastness and the need of engineering and technology in modern civilization will surely open up new windows of innovation in environmental remediation science [2]. Nanotechnology is a wonder of science today. Safety, toxicity, and environmental impact of NMs stand as a crucial pillar of this treatise [2]. There is an immediate need of water purification technologies to remove the micropollutants and intensify the industrial processes. The vision of environmental protection will be widened and the scientific sagacity and doctrine will surely usher in a new age in science and technology [2].

Werkneh et al. [3] discussed with deep scientific insight applications of nanotechnology and biotechnology for sustainable water and wastewater treatment. Environmental or green sustainability is the CoinWord of today's advancement in science and engineering [3]. Nowadays, water pollution and drinking water scarcity have become a serious problem worldwide, causing vast concerns to both public health engineering and environmental engineering science. To reduce these scientific challenges, various innovations and treatment techniques have been envisioned. The vision and sagacity of science and engineering are today in the path of new regeneration. Environmental protection science is today baffled as global warming and global climate change destroys the scientific regeneration. Among these environmental engineering tools, nanotechnology, and biotechnology-based tools are usually applied separately for water and wastewater purposes [3]. This chapter focuses on new and emerging nano- and biotechnologies for the sustainable removal of pollution causing contaminants during water and wastewater treatment strategies. Besides the authors also dealt with toxicological and safety, aspects of different nanotechnologies and their current and future perspectives and recommendations are described. Sustainable water and wastewater treatment tools, nanotechnology perspectives, NMs for the disinfection of pathogenic microbes, and applications of NMs as adsorbent are discussed with lucid insight [3]. Green sustainability is today in the path of newer rejuvenation as civilization treads forward. Environmental biotechnology perspectives and bioremediation techniques are the other cornerstones of this research pursuit. Bioreactor configurations in water and wastewater treatment and a case study in water and wastewater treatment stands as major pillars of this chapter [3]. Today toxicological perspectives of nano/biotechnology are the midst of vast scientific contemplation. The challenge and the vision of environmental remediation are today in the hands of environmental and chemical engineers [3]. A deep thought on bioreactor configurations will surely open new doors of innovation and scientific intuitiveness in years to come. The need of environmental nanotechnology and its vast vision are also elucidated in this chapter [3].

Santhosh et al. [4] deeply comprehended with insight and vision the role of NMs in water treatment applications. Water pollution by various toxic contaminants has become a major issue worldwide [4]. The vision and the future recommendations of this study surpass vast and versatile scientific frontiers. Various technologies and innovations have been used to treat water and wastewater including chemical precipitation, ion-exchange, adsorption, membrane filtration, coagulation-flocculation, flotation, and electrochemical methods. Today nanotechnology has gained immense attention and various NMs have been developed for the water remediation. The immense emphasis has been given to adsorption, photocatalytic, and antibacterial activity of NMs [4]. The authors deeply discussed in this chapter NMs as adsorbents for water treatment, carbon-based materials, application of carbon nanotubes, graphene (GR) based materials, applications of metal oxide-based NMs, miscellaneous nanoadsorbents, NMs as photocatalysts, and the vast relevant domain of NMs as anti-bacterial agents [4]. NMs and engineered NMs are the veritable wonders of science and engineering today. The challenges, the vision, and the ingenuity in NMs applications in environmental remediation are elucidated in minute details in this chapter [4]. Clean water is one of the vital elements for all living organisms to sustain life on earth. Due to rapid industrialization and ever-growing urbanization, the contamination of water resources has occurred immensely in the global scenario today [4]. The validation of science and engineering of environmental protection and the needs of civilization are the path-breakers towards a new era in the field of environmental engineering, chemical process engineering, and water resource management. Some of the important classes of aquatic pollutants are heavy metals and dyes, and once they enter into the water, water is no longer safe for drinking purpose and sometimes it is extremely difficult to treat the contaminated water. For the past few decades, various tools have been developed for treating wastewater. Among the most important methods are solvent extraction, micro, and ultrafiltration, sedimentation, and gravity separation, flotation, precipitation, coagulation, oxidation, evaporation, distillation, reverse osmosis, adsorption, ion exchange, electrodialysis, electrolysis, etc. [4] Technology, engineering, and science thus needs to be re-envisioned and reorganized with the process of civilization and academics. Scientific humanisms and deep scientific and engineering transcendence are the necessities of research pursuit today. This research effort gleans and describes in minute details the veritable success of sustainable environmental engineering processes. Adsorption can be applied for the removal of soluble and insoluble organic, inorganic, and biological pollutants. Technological redeeming and deep engineering emancipation in the field of environmental science and nanotechnology are the needs of the hour. Although many excellent review articles have been published throughout the world discussing the importance of NMs in water treatment and environmental remediation but some of them are only a material and adsorbent-specific (examples: carbon nanotubes, GR based NMs, nanometal oxides, nano zerovalent ion, cellulose NMs) or an adsorbent specific(examples: metals, dyes, pharmaceuticals, and personal care products) [4]. Nanoabsorbents are NS materials from organic or inorganic materials that have high affinity to adsorb contaminants in water [4]. Human society is thus in the path of new scientific regeneration. The special emphasis of this review has been given on adsorption, photocatalytic, and anti-microbial properties of NMs. This chapter widens the scientific thoughts and the future recommendations in the field of application of NMs.

The domain of environmental remediation is today linked with the vast domain of nanotechnology today. Humanity and engineering science are today in the similar manner in the path of new scientific rejuvenation. In this entire chapter, the author pointedly focuses on the success of the science of nanotechnology in the true emancipation of science and technology globally today. The wonders and the ingenuity of science and engineering will surely open new avenues of scientific humanism and deep scientific redeeming in decades to come [4].

10.9.1 RECENT SCIENTIFIC RESEARCH PURSUIT IN THE FIELD OF WATER QUALITY MANAGEMENT

The challenges and the vision to excel in the field of water quality management have no bounds as mankind treads forward. Technological and engineering validation is the needs of scientific progress and the scientific rigor in the field of water quality management. Today's world is a technology driven society. Every avenues of scientific endeavor are linked with engineering and technology. Water quality management is also linked with innovations of engineering science such as NMs and nanotechnology. Integrated water resource management and water quality management are today linked by an unsevered umbilical cord. Industrialization and urbanization are today destroying the scientific firmament of global civilization. In this section, the author deeply comprehends the need of water quality management in the progress of science and civilization. The world of science and engineering of water resource management is also a relevant avenue of research pursuit today.

Food and Agriculture Organization Workshop, Bangkok [5] discussed in an extremely positive manner and scientific foresight water quality management and control of water pollution. The keynote chapters are: (1) water quality management in Asia and Pacific, [5] (2) use of treated wastewater for irrigation, (3) organic sewage treatments, (4) crop production in Bangladesh, (5) water quality and irrigation in China, (6) agriculture and water quality in India towards sustainable management and the deep scientific emancipation of water quality management in South Asia. The expert group emphasized that water is a key resource for all economic sectors which are extremely competitive as a scarce commodity. Global water challenges are in the process of new regeneration. The following are the principal issues as conclusions from the workshop:

1. The need to exploit the potential of treated wastewater [5];
2. Salinity is an important water resource engineering issue;
3. Low water use efficiency;
4. Policy failure;
5. Institutional and legal reform;
6. Data programmes;
7. Management practices;
8. Capacity building [5].

The technology of water quality management and water resource engineering are witnessing drastic changes. Water treatment stands in the middle

Application of Engineered Nanomaterials in Environmental Protection 203

of deep scientific introspection as well as contemplation. This workshop targeted the larger issues confronting the global water challenges.

Biswas et al. [6] deeply discussed with scientific foresight and ingenuity an introductory framework of water quality management. The profundity and the sagacity of the science of water quality management are the pillars of water resource engineering globally today [6]. The present situation due to water scarcity and global warming is getting worse day-by-day. Humanity and civilization need to gear forward towards environmental challenges [6]. The main emphasis in the past and present has been on water quality management including the vast issue of water allocation. From the United States to the United Arab Emirates, and throughout Asia, Africa, and Latin America, water management practices needs space for larger improvement [6]. Sufficient knowledge, technology, and management practices are the veritable pillars of water resource management and water resource engineering. Worldwide, the situation is significantly worse with respect to non-point industrial pollution. According to the Millennium Development Goals water, sufficiency challenges are the pivots of civilization's progress today [6]. In developed countries around the world, a determined long-term effort is highly necessary to successfully control the non-point sources so the ambient quality standards are to be effectively maintained. This intense scientific effort will widen the scientific ingenuity and the future scientific thoughts in the field of water quality management [6].

Ozkal et al. [7] briefly discussed nanoparticles for photocatalytic removal and control of antibiotics, antibiotic resistance, and gene transfer. This technical note presents the trends and fundamentals in the removal and control of antibiotics, antibiotic-resistant bacteria, and antibiotics resistant genes by nanoparticle-based photocatalytic applications [7]. The chapter discussed the mechanism of photoactivation and photocatalysis. Photoactivation of nanoparticles adds veritably more than a single dimension into interaction between photocatalyst and the target pollutants [7]. In the concluding portion of this chapter, efficiency determination of antibiotic resistance control by proposed treatment technologies need proper description of the obtained results in the vast conformity with the terminology. This is just a single case-study of water pollution and its abatement. The science of environmental protection and water treatment needs to be envisioned and reorganized as mankind trudges forward. Humanism and scientific candor are the pillars of research pursuit globally today. In a similar vision, industrial pollution control and water purification are the scientific imperatives of global environmental engineering order. This chapter is a vivid example of the success of environmental engineering science and applications of nanoparticles [7].

Technology transcendence and scientific and engineering revival are the necessities of innovation and intuitiveness in present-day human mankind. Validation of environmental protection science and water quality management will surely open new doors of vision and ingenuity in years to come.

10.9.2 RECENT SCIENTIFIC ENDEAVOR IN THE FIELD OF INTEGRATED WATER RESOURCE MANAGEMENT

Integrated water resource management and urban and rural water quality management are the imminent needs of science and civilization today. Engineering and technology truly has no answers to the ever-growing concerns for water quality and the issue of provision of pure drinking water. Thus, civilization is in deep peril today. The sagacity of environmental protection science and the vision behind it are vast and versatile. Integrated water resource management stands in the middle of deep introspection and scientific profundity. The abundance and transcendence of science and engineering will surely open new windows of innovation in integrated water resource management. Integrated water resource management and water quality management are today connected by an undivided umbilical cord. Sustainable water management is the other visionary avenue of research pursuit.

United Nations Environment Programme Report [8] discussed with lucidity and insight integrated water resource management. This well-researched treatise targets (1) water management and the 2030 agenda, (2) monitoring integrated water resource management in the Sustainable Development Goals, (3) status of integrated water resource management implementation, (4) towards full implementation of integrated water resource management, and (5) greater emancipation of water resource engineering. Water is the lifeblood of global ecosystems and the futuristic vision of human civilization. Water is a condition for global economic prosperity [8]. The 2030 Sustainable Development Goals and the success of its implementation are the main pillars of this report. Sustainable development and water management are the global scientific imperatives. Integrated water resource management is a process which promotes the coordinated development and management of water, land, and related resources. The scientific and technological challenges behind water resource management are deeply elaborated in this report.

Council on Energy, Environment, and Water, India Report [9] deeply discussed with immense vision and farsightedness urban water and sanitation in India. Contemporary India faces a vicious developmental challenge, mainly providing safe and affordable drinking water and proper sanitation to its burgeoning population [9]. Inadequate water supply and sanitation services impose an immense burden to the poor. Sustainable water resource management is thus the vital need. This report delineates:

1. Characteristics of water quality management;
2. Water quality management in the urban sector;
3. Building capacity and an efficient funding mechanism;
4. Planning design strategies in water;
5. Using unconventional options for building efficient urban water sector in India;
6. Private sector participation in water management;
7. Regulatory framework for urban water management in India [9];
8. Water data and measurement;
9. Building water capacity in the urban management scenario [9].

The world of science and engineering of environmental protection and water pollution control needs a thorough revamping. Water quality management in the urban scenario will be a vital area of research pursuit today. The application of technology management and reliability engineering will enhance the global water scenario. Industrial and systems engineering are the other areas of intense areas of research endeavor today. In this chapter, the author rigorously pursues the ingenuity and the technological profundity of global water purification and its scientific innovations such as NMs application. A new dawn of civilization will surely evolve and global water challenges will be surely mitigated.

10.10 ENGINEERED NANOMATERIALS (NMS) AND ENVIRONMENTAL PROTECTION

Environmental remediation and environmental science are the targets of scientific progress today. Millions of citizens around the world are today without pure drinking water. The technological challenges and the truth of science and engineering needs to revisit itself as research and development initiatives in environmental protection gears forward. Human scientific transcendence and deep scientific honing are the needs of science and engineering today. Industrialization and massive pollution in developing and developed nations around the world are mesmerizing science and technology today. Academic and scientific rigor in the field of process engineering and technology management are really changing the face of science and engineering today. Engineered NMs today should be veritably honed in its applications in environmental protection and water/wastewater treatment. Research pursuit in the field of nanotechnology and process engineering are in the forefront of humanity today. Human factor engineering, industrial, and systems engineering and integrated water resource management is the needs of science and engineering today. In the global scenario today, integrated water resource management and wastewater management are today aligned with the vast world of industrial and systems management. Environmental engineering science industry is a big research industry today globally. NMs and engineered NMs are thus in the path of new vision and new regeneration. Technology management is also thus integrated with integrated water resource management. The scientific truth and the vast scientific imagination thus need to be reenvisioned as regards application of nanotechnology in environmental remediation. The author in this chapter stresses a whole lot of application areas of environmental engineering which includes human factor engineering and technology management. Thus, mankind and man's deep vision will be enshrined in the altars of integrated water resource management and nanotechnology. The scientific subtleties of environmental protection will thus be envisioned [10–16].

10.11 FUTURE SCIENTIFIC THOUGHTS AND FUTURE RECOMMENDATIONS OF THIS STUDY

Science and technology in the global scenario are in the path of new rejuvenation. Future of science and engineering thus needs to be re-envisioned and re-vitalized with the march of civilization. Water purification, drinking water treatment, and industrial wastewater treatment today stands in the middle of vast scientific transcendence and scientific contemplation. Industrialization and the advancement of human civilization have veritably destroyed the environment and ecological biodiversity. Human scientific forays are thus in the middle of deep scientific introspection and scientific honing. Developed and developing countries around the world are today faced with world's largest environmental disasters-arsenic groundwater contamination

Application of Engineered Nanomaterials in Environmental Protection 207

in Bangladesh and West Bengal, India. Here come the immediate scientific research and development initiatives in water purification and environmental remediation. Future recommendations of this study and future scientific thoughts today should be targeted towards a greater scientific emancipation in the field of environmental engineering and process engineering fundamentals. Reliability engineering, technology management and human factor engineering are today the needs of scientific progress globally. Future flow of thoughts should be veritably the alignment of nanotechnology with environmental protection and chemical process engineering. Science and technology are today retrogressive as regards global water scarcity and global climate change. Maintenance of ecological biodiversity is the other side of the visionary coin. Here also comes the immense importance of new innovations in nanotechnology. Scientific doctrine, scientific redeeming, and the vast world of scientific intricacies in environmental remediation will eventually lead to a new global environmental engineering order. The crisis in Bangladesh and India are immense and scientifically thought-provoking. This chapter rigorously points towards the truth and sagacity of the science of nanotechnology and environmental remediation. Future flow of scientific thoughts should be reorganized towards innovation, inventions, and ingenuity in the application of nanotechnology and engineered NMs in water and wastewater treatment. Millions of people around the world are immensely in the midst of water crisis and lack of pure drinking water. This chapter with vast academic rigor focuses on the success of nanoadsorbents, nanomembranes, GR nanotubes, and fullerenes in water and wastewater treatment. Human civilization and the progress of science will surely revisit and redeem itself in the field of provision of clean drinking water [21–26]. The marvels of nanotechnology such as NMs and engineered NMs will thus open up new research and development avenues in decades to come [17–20].

10.12   CONCLUSION, SUMMARY, AND SCIENTIFIC PERSPECTIVES

Environmental engineering science, chemical process engineering, and nanotechnology are today in the path of new scientific excellence and deep scientific redeeming. The perspectives of environmental remediation need to be re-envisioned and reorganized as science and technology moves forward. Millions of people around the world are without clean drinking water. Here comes the immediate need of innovations, scientific excellence, and deep scientific instinct. Man's vast scientific re-envisioning, mankind's knowledge prowess, and the needs of human scientific progress will surely invigorate the areas of environmental remediation and environmental sustainability. The global water crisis is devastating today. There are hardly any answers to the ever-growing concerns for heavy metal and arsenic contamination of drinking water globally. Thus the need of an immense scientific research pursuit in water purification, nanotechnology, engineered NMs and integrated water resource management. Human factor engineering, technology management and reliability engineering are the pillars of research pursuit in engineering and technology today. This chapter will surely open new avenues of scientific understanding and scientific redeeming in the field of water purification and nanotechnology in decades to come. Thus, science and technology of engineered NMs will usher in a new era and scientific vision in its applications will open new doors of innovation and instinct. Application of engineered NMs in environmental protection is today reaching new heights and is in the path of new scientific regeneration and vast contemplation. Today science and technology in developing and developed nations around the world are in the similar vein in the process of scientific regeneration. This chapter unfolds the needs of science and engineering of nanotechnology in the scientific progress of human society. Truly, a new visionary era will emerge in the field of science and engineering.

ACKNOWLEDGMENT

The author gratefully acknowledges the contribution of late Shri Subimal Palit, an eminent textile engineer and author's father who taught the author rudiments of chemical engineering.

KEYWORDS

- environmental engineering
- environmental sustainability
- integrated water resource management
- nanomaterials
- nanotechnology
- remediation
- water quality management

Application of Engineered Nanomaterials in Environmental Protection   209

REFERENCES

1. Inter-American Development Bank Report, (2019). *The Future of Water: A Collection of Essays on "Disruptive Technologies that May Transform the Water Sector in the Next 10 Years."* Water and Sanitation Division, Discussion Paper Number-IDB-DP-657.
2. Kunduru, K. R., Nazarkovsky, M., Farah, S., Pawar, R. P., Basu, A., & Domb, A. J., (2017). Nanotechnology for water purification: applications of nanotechnology methods in wastewater treatment. *Chapter 2: Book-Water Purification* (pp. 33–74). Elsevier/Academic Press, New York, USA, Editor-Alex Grumezescu.
3. Werkneh, A. A., & Rene, E. R., (2019). Applications of nanotechnology and biotechnology for sustainable water and wastewater treatment. In: Bui, X. T., et al., (eds.), *Chapter 19: Water and Wastewater Treatment Technologies, Energy, Environment and Sustainability* (pp. 405–430). Springer Nature Singapore Pte Limited-2010.
4. Santhosh, C., Velmurugan, V., Jacob, G., Jeong, S. K., Grace, A. N., & Bhatnagar, A., (2016). Role of nanomaterials in water treatment applications: A review. *Chemical Engineering Journal*, 1–73. doi: http://dx.doi.org/10.1016/j.cej.2016.08.053.
5. Food and Agriculture Organization Workshop, (1999), Water quality management and control of water pollution. *Proceedings of a Regional Workshop.* Bangkok, Thailand.
6. Biswas, A. K., & Tortajada, C., (2011). Water quality management: An introductory framework. *International Journal of Water Resource Management, 27*(1), 5–11.
7. Ozkal, C. B., Brienza, M., & Meric, S., (2019). Nanoparticles for photocatalytic removal and control of antibiotics, antibiotic resistance, and resistance gene transfer. *Journal of Desalination and Water Purification, 14*, 3–6.
8. United Nations Environment Programme Report, (2018). *Progress on Integrated Water Resource Management*. Degree of IWRM Implementation.
9. Council on Energy, Environment and Water, India Report, (2013). *Urban Water and Sanitation in India: Multi-Stakeholder Dialogues for Systemic Solutions.*
10. Palit, S., (2015). Microfiltration, groundwater remediation, and environmental engineering science-a scientific perspective and a far-reaching review. *Nature, Environment and Pollution Technology, 14*(4), 817–825.
11. Palit, S., & Hussain, C. M., (2018). Biopolymers, nanocomposites, and environmental protection: A far-reaching review. In: Shakeel, A., (ed.), *Bio-Based Materials for Food Packaging* (pp. 217–236). Springer Nature Singapore. Pte Ltd.
12. Palit, S., & Hussain, C. M., (2018). Nanocomposites in packaging: A groundbreaking review and a vision for the future In: Shakeel, A., (ed.), *Bio-Based Materials for Food Packaging* (pp. 287–303). Springer Nature Singapore. Pte Ltd.
13. Palit, S., (2017). Advanced environmental engineering separation processes, environmental analysis and application of nanotechnology-a far-reaching review. Hussain, C. M., & Kharisov, B., (eds.), *Chapter-14: Advanced Environmental Analysis-Application of Nanomaterials* (Vol. 1, pp. 377–416). The Royal Society of Chemistry, Cambridge, United Kingdom.
14. Hussain, C. M., & Kharisov, B., (2017). *Advanced Environmental Analysis-Application of Nanomaterials* (Vol. 1). The Royal Society of Chemistry. Cambridge, United Kingdom.
15. Hussain, C. M., (2017). Magnetic nanomaterials for environmental analysis. In: Hussain, C. M., & Kharisov, B., (eds.), *Chapter 19: Advanced Environmental Analysis-Application*

of Nanomaterials (Vol. 1, pp. 3–13). The Royal Society of Chemistry, Cambridge, United Kingdom.

16. Hussain, C. M., (2018). *Handbook of Nanomaterials for Industrial Applications.* Elsevier, Amsterdam, Netherlands.

17. Palit, S., & Hussain, C. M., (2018). Environmental management and sustainable development: A vision for the future. In: Chaudhery, M. H., (ed.), *Handbook of Environmental Materials Management* (pp. 1–17). Springer Nature Switzerland AG.

18. Palit, S., & Hussain, C. M., (2018). Nanomembranes for environment. In: Chaudhery, M. H., (ed.), *Handbook of Environmental Materials Management* (pp. 1–24). Springer Nature Switzerland AG.

19. Palit, S., & Hussain, C. M., (2018). Remediation of industrial and automobile exhausts for environmental management. In: Chaudhery, M. H., (ed.), *Book-Handbook of Environmental Materials Management* (pp. 1–17). Springer Nature Switzerland AG.

20. Palit, S., & Hussain, C. M., (2018). Sustainable biomedical waste management. In: Chaudhery, M. H., (ed.), *Handbook of Environmental Materials Management* (pp. 1–23). Springer Nature Switzerland AG.

21. Palit, S., (2018). Industrial vs. food enzymes: Application and future prospects. In: Mohammed, K., (ed.), *Enzymes in Food Technology: Improvements and Innovations* (pp. 319–345). Springer Nature Singapore Pte. Ltd., Singapore.

22. Palit, S., & Hussain, C. M., (2018). Green sustainability, nanotechnology and advanced materials-a critical overview and a vision for the future. In: Shakeel, A., & Chaudhery, M. H., (eds.), *Chapter 1: Green and Sustainable Advanced Materials, Applications* (Vol. 2, pp. 1–18). Wiley Scrivener Publishing, Beverly, Massachusetts, USA.

23. Palit, S., (2018). Recent advances in corrosion science: A critical overview and a deep comprehension. In: Kharisov, B. I., (ed.), *Direct Synthesis of Metal Complexes* (pp. 379–410). Elsevier, Amsterdam, Netherlands.

24. Palit, S., (2017). *Nanomaterials for Industrial Wastewater Treatment and Water Purification.* In: *Handbook of Ecomaterials* (pp. 1–41). Springer International Publishing, AG, Switzerland.

25. www.wikipedia.com (accessed on 16 June 2020).

26. www.google.com (accessed on 16 June 2020).

CHAPTER 11

Chemistry and Sustainable Development

FRANCISCO TORRENS[1] and GLORIA CASTELLANO[2]

[1]*Institute for Molecular Science, University of Valencia, PO Box 22085, E–46071 Valencia, Spain, E-mail: torrens@uv.es*

[2]*Department of Experimental Sciences and Mathematics, Faculty of Veterinary and Experimental Sciences, Valencia Catholic University Saint Vincent Martyr, Guillem de Castro-94, E–46001 Valencia, Spain*

ABSTRACT

The expression relating medicine to war. Ad hoc or unofficial (self-)therapeutic upkeep continues; tobacco, ethanol, and several other painkillers result inherent to war. Several queries were educated on robust connections between soldierly and noncombatant painkiller-getting How must soldier-health establishment understand individual painkiller utilization in battle? Self-therapy? Dogmatic confrontation? A surviving plan? How must they reply? Fissile knowledge innovation compares to the backward-looking philosophy of the Franco scheme. No past association occurs between knowledge and equality. To rebuild the procedures of creating unawareness: the movement of creating unawareness. Peace Boat is a global non-administrative group that functions to endorse concord and maintainability via arrangement of concord journeys aboard a nearside boat. Actions result founded on the attitude that one difficulty confronted by one group is a worldwide test, which should be undertaken via persons-associations-administrations collaboration. It results sensible to take up that spiritual origin, originated from physical states of all civilizations, directed particular persons by the condition of positioning development greed opposite agreement with nature, which produced quick growth of significant ethical constructions that lied in specialized development, nonetheless likewise positioned the fraction of humankind that beard the main position in them in a boundary circumstances.

11.1 INTRODUCTION

Setting the scene: World War I (WWI), health, medicine, humanitarian, monkeyman, poison gas and WWI, the Great War *via* images, the fall of democracy in the Inter-war Period, the power of chemistry in Franco's Spain and prey-predator model reflected in Lotka-Volterra equations to discover the magnetic confinement of plasma. The equation: Medicine = War. *Ad hoc* or informal (self-)medical care persists; tobacco, alcohol, and any other drugs; drugs are intrinsic to warfare. The following questions (Q) were raised on strong links between military and civilian drug-taking:

Q1. How should military-medical authorities see personal drug use in combat?
Q2. Self-medication?
Q3. Political resistance?
Q4. A coping strategy?
Q5. How should they respond?

Nuclear science modernity contrasts with the reactionary ideology of the Franco system. There is no necessary historical relationship between science and democracy. To reconstruct the processes of making of ignorance: the campaign of making of ignorance. Peace Boat is an international non-governmental organization that works to promote peace and sustainability *via* organization of peace voyages onboard a passenger ship. Activities must be based on the philosophy that any problem faced by any community is a global challenge that must be tackled *via* people-organizations-governments co-operation. It is reasonable to assume that religious root, born of material conditions of every society, guided certain people by the way of placing growth avidity in front of harmony with environment, which caused fast development of important philosophical structures that rested on technical advance, but also placed the part of humanity that took the chief role in them in a limit situation, in which its own survival capacity is threatened.

Technoscientific organisms and the history of fascism were reviewed [1]. The conversion of the atom, nuclear sciences, and ideology in the Franco years were informed [2]. The beginning of nuclear science in Spain was revised [3]. American nuclear cover-up in Spain after Palomares (Almería, Spain) disaster (1966) was discussed [4]. Ignorance *vs.* conscience-making in Palomares accident was reported [5]. A nuclear weapons *radiography* was published [6]. It was examined Chernobyl, Fukushima, and Cofrents

Chemistry and Sustainable Development

(València, Spain) nuclear power station of electric generation type boiling water reactor (BWR)-6 (1984) [7].

In earlier publications, it was informed nuclear fusion, American nuclear cover-up in Spain after Palomares (Almería, Spain) disaster (1966) [8], *Manhattan Project, Atoms for Peace*, nuclear weapons, accidents [9], nuclear science, technology [10], history, concept, method, didactics of atomic and nuclear physics [11], gravitational waves, messengers of the universe, cosmology [12], plasma, photo/radiochemical reactions and relativity theories [13]. The aim of the present work is to review WWI, health, medicine, humanitarian, monkeyman, poison gas and WWI, the Great War *via* images, the fall of democracy in the Inter-war Period, the power of chemistry in Franco's Spain and prey-predator model reflected in Lotka-Volterra equations to discover the magnetic confinement of plasma. The goal of this work is to initiate a debate by suggesting a number of questions on chemical weapons, treatment of injured, humanitarian challenges, modern medicine, old-fashioned drink, and drugs, coping with wounds, disease, and trauma on Western Front, and the power of chemistry in Franco's Spain, and providing, when possible, answers (A) and hypotheses (H).

11.2 MONKEYMAN: POISON GAS AND WORLD WAR I (WWI)

In a course of seminars *In the Centenary of the Armistice: Health, Medicine, and Humanitarian Immediately after WWI (1914–1918)*, a hypothesis was proposed.

H1.   The equation: Medicine = War.

Van Bergen raised Q on chemical weapons, treatment of injured and humanitarian challenges (*cf.* Figure 11.1) [14]:

Q1.   Which were the real mortality rates [15]?
Q2.   When could one affirm that a person was poisoned?
Q3.   Which were the pathological consequences of the poisonings?
Q4.   What could the physicians and nurses do for the victims?
Q5.   Which psychological effects do gas masks cause?

Gorin discussed visual representations of the medical and humanitarian assistance to the wounded and to the invalids during WWI (*cf.* Figure 11.2) [16].

FIGURE 11.1 English soccer team at North France (1916).

FIGURE 11.2 WWI, Geneva, Repatriation of gravely wounded French soldiers, arrested in Germany, while being assisted by health personnel of Swiss Red Cross in a station.

Reid raised Q on modern medicine, old-fashioned drink and drugs, and coping with wounds, disease, and trauma on Western Front [17]:

Q6. Till which point self-medication is useful to help men to tackle conflict demands [18–20]?
Q7. In which point self-medication turns destructive?
Q8. Which role should play the army in the monitoring and control of the abuse of substances?

She provided the following conclusions (Cs):

C1. *Ad hoc* or informal (self-)medical care persists.
C2. Tobacco, alcohol, and any other drugs.
C3. Drugs are intrinsic to warfare.
C4. Strong links between military and civilian drug-taking.
Q9. How should military-medical authorities see personal drug use in combat?
Q10. Self-medication?
Q11. Political resistance?
Q12. A coping strategy?
Q13. How should they respond?

Borowi presented disaster, co-operation, and paradoxical WWI role on international public health (*cf.* Figure 11.3) [21–24].

11.3 THE GREAT WAR VIA IMAGES

The WWI as a war conflict took place in 1914–1918. It produced the confrontation between the main imperialist powers, especially European ones (e.g., the Triple Entente, Triple Alliance) who competed not only in the economic and political sphere but also in the technologies and the development of armament, which allowed imposing themselves before the others and prepare for the confrontations in a more effective way. The outbreak of WWI was apparently enthusiasm, produced by the fact that presenting so much military technology, it would end in a short time, which ironically gave rise to the development, creation, and refinement of new weapons, which were responsible for causing much more destruction within the territories. Arias Rodriguez reported the history of WWI and the arms race that originated [25]. It was focused on a series of images of the time that relate the main advances of weapons, their characteristics, and the specific function they fulfilled on the battlefield. The technological and industrial development did not stop after the end of the war, and what is intended is to demonstrate the impact generated by these new technologies in the course of the war.

FIGURE 11.3 From disaster to co-operation: The paradoxical role of WWI on international public health.

11.4 THE FALL OF DEMOCRACY IN THE INTER-WAR PERIOD

The fall of democracy was a political phenomenon that happened after the fall of the empires at the end of WWI. Vélez analyzed the phenomenon since Bolshevik Revolution, taking it as a cause not only of democratic intervention in Russia but also as a reaction action for the emergence of other totalitarian movements that, although different, did not allow the development of a parliamentary policy in different countries [26]. The struggle to maintain democracy always generated internal civil wars that conceived certain social dynamics (e.g., social segregation, class struggle, strong nationalism).

Chemistry and Sustainable Development

Although in Germany it was constituted during a good part of the Inter-war Period as a democratic republic, the division and fragmentation presented by the German democratic movement prevented the development of democracy in Germany, which Hitler took advantage of to take power.

11.5 THE POWER OF CHEMISTRY IN FRANCO'S SPAIN

Nieto-Galan proposed H/Q/A on the power of chemistry in Franco's Spain (*cf.* Figure 11.4) [27]:

H1. New (social, etc.), ethos: Spain–US Agreement (Pact of Madrid, 1953).

H2. (M. Lora-Tamayo). Better things for better living through chemistry.

H3. Not all fascist science is bad but excellent science requires social freedom (imagination, etc.).

Q1. (I. Monzó). Why do not chemistry professors explain the sociology of chemistry?

A1. There are two hypotheses.

H4. Practically, no historical rhetoric exists and, in Spanish-Civil-War Generation, totals rupture.

H5. Ingenuity/particular nature of chemistry for armor-plating itself (compare it with Darwinism).

H6. Tacit knowledge: intangible.

Q2. Why progress a determined reaction occurs or does not?

A2. Any science requires basic science, at least to training researchers.

H7. Operation Paperclip: many German scientists recruited in post-Nazi Germany and taken to the US for government employment.

Q3. (X. Guillem). In Operation Paperclip, did these German scientists pass to Spain?

A3. Work on free radicals (FRs) continues EidgenössischeTechnische Hochschule (ETH) Zürich in 1910.

Q4. (X. Guillem). Were different research lines because of ideology?

A4. Yes, carbon derivatives for fuel because there was no money for oil.

Q5. (I. Monzó). Is it necessary for a scientist to do a stay abroad?

A5. A net is not to do a stay (asymmetric net) but bring back the head for a stay (symmetric net).

H8. There are historical processes that require a long time.

FIGURE 11.4 The power of chemistry in Franco's Spain.

11.6 PREY-PREDATOR MODEL TO DISCOVER THE MAGNETIC CONFINEMENT OF PLASMA

Keeping running, the present world just as people know it requires the production of a huge amount of energy [28]. Although nuclear fusion could

Chemistry and Sustainable Development

answer the necessity, the development of a technology that allows its use is not exempt from complications. On one hand, to keep the plasma in magnetic confinement at high temperatures (HTs) is necessary. On the other hand, difficulties exist in knowing the plasma behavior in the confinement because of fluctuations and turbulences, in which sense, it is particularly interesting using simple models that describe the main characteristics of the turbulence dynamics, e.g., prey-predator model reflected in Lotka-Volterra equations.

11.7 FINAL REMARKS

From the present results and discussion, the following final remarks can be drawn:

1. The equation: Medicine = War.
2. *Ad hoc* or informal (self-)medical care persists.
3. Tobacco, alcohol, and any other drugs.
4. Drugs are intrinsic to warfare.
5. Strong links between military and civilian drug-taking.
6. How should military-medical authorities see personal drug use in combat? Self-medication? Political resistance? A coping strategy? How should they respond?
7. Nuclear science modernity contrasts with the reactionary ideology of the Franco system. No necessary historical science-democracy relationship exists. American nuclear cover-up in Spain after Palomares disaster showed making of ignorance *vs.* conscience *via* official *vs.* local speeches. The objective is to reconstruct the processes of making of ignorance *via* its campaign.
8. Peace Boat is an international non-governmental organization that works to promote peace and sustainability *via* organization of peace voyages onboard a passenger ship. Activities must be based on the philosophy that any problem faced by any community is a global challenge that must be tackled *via* people-organizations-governments co-operation.
9. It is reasonable to assume that religious root, born of material conditions of every society, guided certain people by the way of placing growth avidity in front of harmony with environment, which caused fast development of important philosophical structures that rested on technical advance, but also placed the part of humanity that took the chief role in them in a limit situation.

10. *Sustainable development* concept connects directly with West anthropocentric cultural tradition. An achievement is emphasizing, on raising them to the level of international political discussion, *synchronic* and *diachronic solidarity* concepts. Sustainable development needs to understand that Earth is a finite system, with a limited external power contribution.

ACKNOWLEDGMENTS

The authors thank support from Fundacion Universidad Catolica de Valencia San Vicente Martir (Project No. 2019-217-001UCV).

KEYWORDS

- **boiling water reactor**
- **confrontation**
- **dictatorship**
- **fascism**
- **monkeyman**
- **Nazism**
- **World War I**

REFERENCES

1. Saraiva, T., (2016). *Fascist Pigs: Technoscientific Organisms and the History of Fascism*. Massachusetts Institute of Technology: Cambridge, MA.
2. Roqué, X., (2016). Nuclear sciences and ideology in the Franco regime. *Mètode, 2016*(90), 77–83.
3. Soler, F. P., (2017). The Beginning of Nuclear Science in Spain. Sociedad Nuclear *Española: Madrid*, Spain.
4. Howard, J., (2016). *White Sepulchres: Palomares Disaster Semicentennial Publication.* University of Valencia: Valencia, Spain.
5. Florensa, C., (2018). *Book of Abstracts, Science, Politics, Activism, and Citizenship* (pp. 0–21). València, Spain. REDES CTS-Catalan Society for the History of Science and Technics: València, Spain.
6. Book of Abstracts, (2017). *Nuclear Weapons Radiography*. València, Spain. Peace Studies Delàs Center: València, Spain.

Chemistry and Sustainable Development

7. Montón, R., & Hernàndez, F. J., (2017). *Chernobyl, Fukushima and the Cofrentes Nuclear Power Plant*. Debats No. 31, InstitucióAlfons el Magnànim–CVEI–Diputació de València: València, Spain.

8. Torrens, F., & Castellano, G., (2019). Nuclear fusion and the American nuclear cover-up in Spain: Palomares disaster (1966). In: Haghi, R., & Torrens, F., (eds.), *Engineering Technology and Industrial Chemistry with Applications* (pp. 297–308.). Apple Academic-CRC: Waretown, NJ.

9. Torrens, F., & Castellano, G., (2020). Manhattan project, atoms for peace, nuclear weapons, and accidents. In: Pogliani, L., Torrens, F., & Haghi, A. K., (eds.), *Molecular Chemistry and Biomolecular Engineering: Integrating Theory and Research with Practice* (pp. 215–233). Apple Academic-CRC: Waretown, NJ.

10. Torrens, F., & Castellano, G., (2010). Nuclear science and technology. In: Pogliani, L., Ameta, S. C., & Haghi, A. K., (eds.), *Chemistry and Industrial Techniques for Chemical Engineers*. Apple Academic-CRC: Waretown, NJ, pp. 311–330. in press.

11. Torrens, F., & Castellano, G. History, concept, method, and didactics of atomic/nuclear physics. In: Esteso, M. A., Ribeiro, A. C. F., & Haghi, A. K., (eds.), *Physical Chemistry and Its Interdisciplinary Applications*. Apple Academic-CRC: Waretown, NJ, in press.

12. Torrens, F., & Castellano, G. Gravitational waves, messengers of the universe, and cosmology. In: Esteso, M. A., Ribeiro, A. C. F., & Haghi, A. K., (eds.), *Chemistry and Chemical Engineering for Sustainable Development: Best Practices and Research Directions*. Apple Academic-CRC: Waretown, NJ, in press.

13. Torrens, F., & Castellano, G. Plasma, photo/radiochemical reactions and relativity theories. In: Kulkarni, S., Rawat, N. K., & Haghi, A. K., (eds.), *Green Chemistry and Green Engineering: Processing, Technologies, Properties, and Applications.*. Apple Academic-CRC: Waretown, NJ, in press.

14. Van, B. L., (2019). *Personal Communication*.

15. Van, B. L., (2016). *Before my Helpless Sight: Suffering, Dying and Military Medicine on the Western Front, 1914–1918*. Routledge; Abingdon, UK.

16. Gorin, V., (2019). *Personal Communication*.

17. Reid, F., (2019). *Personal Communication*.

18. Reid, F., (2010). *Broken Men: Shell Shock, Treatment, and Recovery in Britain 1914–1930*. Continuum: London, UK.

19. Gemie, S., Humbert, L., & Reid, F., (2011). *Outcast Europe: Refugees and Relief Workers in an Era of Total War 1936–1948*. Continuum: London, UK.

20. Reid, R., (2017). *Medicine in First World War Europe: Soldiers, Medics, Pacifists*. Bloomsbury: London, UK.

21. Borowi, I., (2019). *Personal Communication*.

22. Borowy, I., & Gruner, W., (2005). *Facing Illness in Troubled Times: Health in Europe in the Interwar Years, 1918–1939*. Peter Lang: Frankfurt am Main, Germany.

23. Borowy, I., (2009). *Coming to Terms with World Health: The League of Nations Health Organization 1921–1946*. Peter Lang: Frankfurt am Main, Germany.

24. Borowy, I., (2014). *Defining Sustainable Development for Our Common Future: A History of the World Commission on Environment and Development (Brundtland Commission)*. Routledge: Abingdon, Oxon, UK.

25. Arias, R. D. L., (2020). *Book of Abstracts* (PTT–20). XXVII Congreso Internacional de Aprendizaje, València, Spain, Red de Investigación de Aprendizaje: València, Spain.

26. Vélez, F., (2020). *Book of Abstracts* (PTT–26). XXVII Congreso Internacional de Aprendizaje, València, Spain. Red de Investigación de Aprendizaje: València, Spain.
27. Nieto-Galan, A. *La Químicaen la España de Franco*; work in preparation.
28. Jiménez-Cómez, M., (2019). *Personal Communication.*

CHAPTER 12

Manganese Porphyrins as Pro-Oxidants in High-Molar-Mass Hyaluronan Oxidative Degradation

KATARÍNA VALACHOVÁ, [1] PETER RAPTA, [2] INES BATINIC-HABERLE, [3] and LADISLAV ŠOLTÉS[1]

[1] *Center of Experimental Medicine, Institute of Experimental Pharmacology and Toxicology, Slovak Academy of Sciences, Bratislava, Slovakia, E-mail: katarina.valachova@savba.sk (K. Valachová)*

[2] *Institute of Physical Chemistry and Chemical Physics, Slovak University of Technology in Bratislava, Bratislava, Slovakia*

[3] *Ines Batinic-Haberle, Department of Radiation Oncology-Cancer Biology, Duke University Medical Center, Durham, USA*

ABSTRACT

The ability of manganese ions and Mn-porphyrins as potential pro-oxidants were examined in ascorbate-induced hyaluronan (HA) degradation by means of rotational viscometry. Further, their effect was examined in HA degradation initiated by Cu(II) ions in the presence of ascorbic acid (AA) as a source of ˙OH radicals and alkoxy-/peroxy-type radicals. Addition of Mn(II) ions or Mn-porphyrins resulted in ascorbate-induced HA degradation. However, after the addition of Cu(II) ions to the HA solution with AA and Mn(II) ions or Mn-porphyrins the degradation of HA was promoted in a dose-dependent manner. HA degradation induced by Cu(II) ions and AA in the presence of Mn(II) ions was a bit slower than in the presence of Mn-porphyrins. Production of ˙OH radicals in the latter mentioned HA oxidative system was detected by electron paramagnetic resonance (EPR).

12.1 INTRODUCTION

Metal ions such as copper, iron, zinc, and manganese are involved in many crucial biological processes and are necessary for the survival of all living organisms. Due to the unique redox potential of some of these transition metals, they serve important roles as cofactors in enzymes. It is estimated that 30–45% of known enzymes are metalloproteins whose functions require a metal co-factor [1–4].

Manganese in a form of Mn^{2+} ions serves as a constituent of metalloenzymes and activators of many metal-enzyme complexes and is found in the diet. Manganese absorption and excretion is tightly controlled to maintain its stable tissue levels for reactions. The most widely known enzyme requiring Mn is manganese superoxide dismutase (MnSOD), whose primary function is detoxification of $O_2^{\cdot-}$ formed within the mitochondria. Mn^{3+} ions are found in the essential enzymes such as catalase and MnSOD. The average person body contains only about 12 mg of manganese, whereas its daily uptake in adults is between 0.7 and 10.9 mg [5, 6]. Mn metalloenzymes including arginase, glutamine synthetase, phosphoenolpyruvate decarboxylase, and MnSOD participate in metabolism of glucose and lipids; acceleration in the synthesis of protein, vitamin C and vitamin B; and the operation of RNA, catalysis of hematopoiesis; regulation of the endocrine; and improvement in immune function, even it maintains the normalization of the synthesis and secretion of insulin and also reduce oxidative stress against free radicals (FRs) [7–9]. Manganese is required for the activation of prolidase, an enzyme that functions to provide the amino acid proline for collagen formation in human skin cells [10]. Glycosaminoglycan synthesis, which requires manganese-activated glycosyl transferases, may also play an important role in wound healing[11].

However, manganese and other transition metals at high intracellular concentrations impact the cellular redox potential and produce reactive oxygen species (ROS) such hydroxyl radicals, which contribute to their toxicity[3]. The valence of manganese can be changed within the body and studies suggest that Mn^{3+} is more cytotoxic than Mn^{2+} due to higher oxidative reactivity, which was supported by a study in rats given either Mn^{2+} or Mn^{3+} ions. Oxidative stress is also significant in Mn-induced dopaminergic neurodegeneration [5, 6]. Also, environmental or occupational manganese overexposure is harmful to human health, especially in at risk populations such as miners, welders, and steel makers[8, 12].

Metalloporphyrins catalyze numerous redox reactions, in particular, manganese porphyrins (MnPs) have been used as redox catalysts in several model systems relevant to biochemistry, for example, as superoxide dismutase and catalase. MnPs have been originally developed as SOD mimics based on a structure activity relationship correlating the metal-centered reduction potential and the rate constant for the catalysis of $O_2^{\cdot-}$ dismutation. MnPs have several redox states and can reduce the levels of reactive species, which results in less damage of biological molecules and in the modulation of transcription factors, protein, and gene expression. The most effective MnPs have been tested *in-vivo*, which showed remarkable efficacy in cardiac, kidney, and liver ischemia, radioprotection, sickle cell disease, tumor suppression, diabetes, and disorders of central nervous system [13]. Porphyrin-based SOD mimics have a redox-active metal (Mn, Fe, and Cu) center and a stable porphyrin complex. The dismutation of $O_2^{\cdot-}$ by Mn-porphyrin complexes involves two steps: in the first step Mn(III) are reduced by $O_2^{\cdot-}$ to yield Mn(II) and O_2 and in the second step Mn(II) are oxidized by $O_2^{\cdot-}$ to yield H_2O_2 and return the manganese to its resting state as Mn(III) porphyrin. However, in the presence of a reductant such as ascorbate, MnPs function as $O_2^{\cdot-}$ reductases rather than dismutases. Mn(III) can be reduced to Mn(II) by ascorbate while Mn(II) can react with O_2 forming $O_2^{\cdot-}$, which subsequently forms H_2O_2 and O_2. The cytotoxic effects of two Mn(III) alkylpyridylporphyrins such as MnTE-2-PyP^{5+} and MnTnHex-2-PyP^{5+} and ascorbate have been demonstrated in Caco-2, HeLa, HCT116, and 4T1 cells. Given that several MnPs have already been tested *in-vivo* as SOD mimetics, and by themselves have shown low toxicities at micromolar levels, there is great potential for using MnPs as an adjuvant to enhance the efficacy of pharmacologic ascorbate [14].

MnPs react with hydroxyl radical, $O_2^{\cdot-}$ and H_2O_2 [Batinic-Haberle and Tome, 2019]. The decrease in oxidative stress injuries in the presence of MnPs have been shown in numerous studies, e.g., radioprotective effects where MnP decreased oxidative damage by scavenging reactive oxygen and nitrogen species (ROS/RNS) or ischemia/reperfusion injuries. High positive oxidation potential of $MnCl_2$ $(+ 850\,mV)$ allows it to act as a reductant only, thus preventing it to oxidize-SH groups. MnP with $E_{1/2} \approx +200\,mV$ can easily donate and accept electrons from redox able compounds such as cellular reductants. As compared to other types of antioxidants, most potent MnPs, particularly, meso substituted MnTE-2-PyP^{5+} and MnTnHex-2-PyP^{5+}, are extremely stable under all conditions of acidity, dilution, temperature, and light [15].

The major *in-vivo* action of MnPs, in the presence of H_2O_2 and glutathione, is most likely the oxidation or S-glutathionylation of cysteines in proteins by a GPx-like manner. In cells, MnPs can act as either reductants or oxidants in reactions with other biologically relevant molecules in a cellular milieu. MnPs can react with biologically relevant sulfur species such as sulfite and hydrogen sulfide. The reduction of Mn(III)P with ascorbate, which results in the formation of ascorbyl radical, may be preferred *in-vivo* [Batinic-Haberle and Tome, 2019].

MnTE-2-PyP^{5+} was the first compound developed. Because of its hydrophilicity, a 5,000-fold more lipophilic analog was developed with lengthened hexyl alkylpyridyl chains known as MnTnHex-2-PyP^{5+} and its higher mitochondrial distribution and transfer across the blood-brain barrier have been demonstrated. Insertion of oxygen atoms into its hydrophobic chains resulted in the synthesis of MnTnBuOE-2-PyP^{5+}, which showed less toxicity than MnTnHex-2-PyP^{5+} whereas high lipophilicity and redox-related performance were maintained [16].

Ferrer-Sueta et al. [13] hypothesized that cationic MnPs could accumulate in mitochondria. In a study by Ferrer-Sueta et al. [17], ≥ 3 μM MnTE-2-PyP^{5+} was able to protect submitochondrial particles against ONOO$^-$ flux. They found that MnTE-2-PyP^{5+} targets mouse heart mitochondria at level 5.1 μM, which can protect against ONOO$^-$-mediated damage. Mouraviev et al. (2011) showed that after a single dose of MnTE-2-PyP^{5+}, contrast changes in prostate tumors were up to six-fold greater than in surrounding, noncancerous tissues. Therefore, they suggested a potential use of this metalloporphyrin as a novel diagnostic probe for detecting prostate malignancy using MRI. MnTnHex-2-PyP^{5+} in combination with ionizing radiation caused a significant delay in the growth of 4T1 and B16 xenograft tumors. MnTnHex-2-PyP^{5+} dose-dependently enhanced ionizing radiation-mediated production of H_2O_2-derived species, but not O_2^- [18] and is distributed at the highest levels to all organs and shows decreased toxicity induced by its micellar character compared to the initial lipophilic analogs. Due to its high bioavailability, MnTnHex-2-PyP^{5+} has a better therapeutic window than MnTE-2-PyP^{5+}.

Most frequently, MnPs have been administered subcutaneously. This method of delivery of MnTE-2-PyP^{5+}, MnTnHex-2-PyP^{5+}, and MnTnBuOE-2-PyP^{5+} results in the distribution to all organs. The more lipophilic MnTnBuOE-2-PyP^{5+} distributes to much higher levels in all organs than does MnTE-2-PyP^{5+}; the difference is the largest with regards to their accumulation in the brain [Batinic-Haberle and Tome, 2019].

Both MnTE-2-PyP^{5+} and MnTnHex-2-PyP^{5+} have protected normal rat lung from radiation-induced oxidative stress. MnTnBuOE-2-PyP^{5+} exhibited

Manganese Porphyrins as Pro-Oxidants

radiosensitizing properties to slow glioblastoma tumor growth in flank tumor growth assays. The addition of MnTnBuOE-2-PyP^{5+} to radiation therapy for brain tumors could be beneficial in preserving healthy brain structure and function while simultaneously antagonizing tumor survival [16].

Evans et al. [19] demonstrated the unique ability of two Mn-based SOD mimics, namely MnTE-2-PyP^{5+} and MnTnBuOE-2-PyP^{5+} to function as pro-oxidants in the presence of ascorbate and enhanced ROS-mediated cell death. The combination of MnP and ascorbate induced accumulation of mitochondrial $O_2^{\cdot-}$- and H_2O_2-derived radicals in both therapy-sensitive and therapy-resistant cell lines.

Hyaluronan (HA) is a high-molar-mass glycosaminoglycan of molar mass several megadaltons composed of repeating disaccharide units consisting of D-glucuronate and N-acetyl-D-glucosamine. It is an important component of most connective tissues, including vitreous body, skin, synovial fluid, and umbilical cord. It affects many cell functions, such as proliferation, differentiation, and migration in a concentration and molar-mass dependent manner [20–22].

HA is susceptible to degradation by ROS, especially by highly reactive ·OH radicals [23, 24], (Andre and Villain, 2016). An efficient ·OH radical generating system is composed of ascorbate and Cu(II) ions and is known as Weissberger's biogenic oxidative system (WBOS, Scheme 1) and the process of HA degradation is denoted in Scheme 2.

The aim of the study was to examine antioxidative or pro-oxidative effects of Mn(II) ions and Mn-porphyrins such as Mn(III)meso-tetrakis(N-ethylpyridinium-2-yl)porphyrin, Mn(III)meso-tetrakis(N-n-hexylpyridinium-2-yl) porphyrin, and Mn(III)meso-tetrakis(N-n-butoxyethylpyridinium-2yl) porphyrintoward the Weissberger biogenic oxidative system-induced degradation of high-molar-mass HA in-$vitro$ (Schemes 1 and 2).

12.2 EXPERIMENTAL PART

12.2.1 MATERIALS

Hyaluronan (M_w = 1.69 MDa) was gained from Lifecore Biomedical Inc., Chaska, MN, USA (content of transition metals: copper < 1 ppm, iron 6 ppm). $CuCl_2 \cdot 2H_2O$ and NaCl p.a. were purchased from Slavus Ltd., Bratislava, Slovakia. $MnCl_2 \cdot 4H_2O$ p.a. was purchased in Sigma-Aldrich, Germany. Ascorbic acid (AA) was purchased in Merck KGaA, Darmstadt, Germany.

Mn(III)meso-tetrakis (*N*-ethylpyridinium-2-yl)porphyrin (MnTE-2-PyP^{5+}), Mn(III) meso-tetrakis(*N*-n-hexylpyridinium-2-yl)porphyrin (MnTnHex-2-PyP^{5+}), and Mn(III)meso-tetrakis(*N*-n-butoxyethylpyridinium-2yl) porphyrin (MnTnBuOE-2-PyP^{5+}) were from Duke University School of Medicine, Durham, USA. 5, 5-Dimethyl-1-pyrroline*N*-oxide, $\geq 97\%$ was purchased in Sigma-Aldrich, Saint Louis, USA. Deionized high-purity grade water, with conductivity of ≤ 0.055 μS/cm, was produced by using the TKA water purification system (Water Purification Systems GmbH, Niederelbert, Germany).

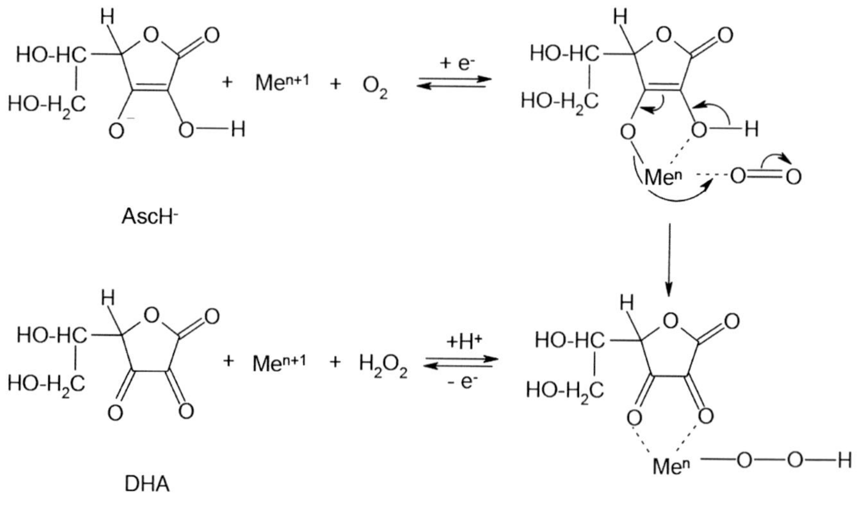

$$H_2O_2 + Me^n\text{-}complex \rightarrow {}^\bullet OH + Me^{n+1} + OH^-$$

SCHEME 1   TheWeissberger biogenic oxidative system, Me-metal.

12.2.2 PREPARATION OF STOCK AND WORKING SOLUTIONS

The HA samples (1.75 mg/ml) were dissolved in 0.15 M aqueous NaCl solution for 24 h in the dark. HA sample solutions were prepared in two steps: first, 4.0 ml of 0.15 M NaCl was added to HA to swell and after 6 h, 0.15 M NaCl in the volumes from 3.9 to 3.76 ml was added, when working in the absence and presence of Mn(II) ions or MnPs. Solutions of AA (16 mM) and cupric chloride (160 μM) were made in 0.15 M aqueous NaCl. Solutions of MnPs were made in deionized water.

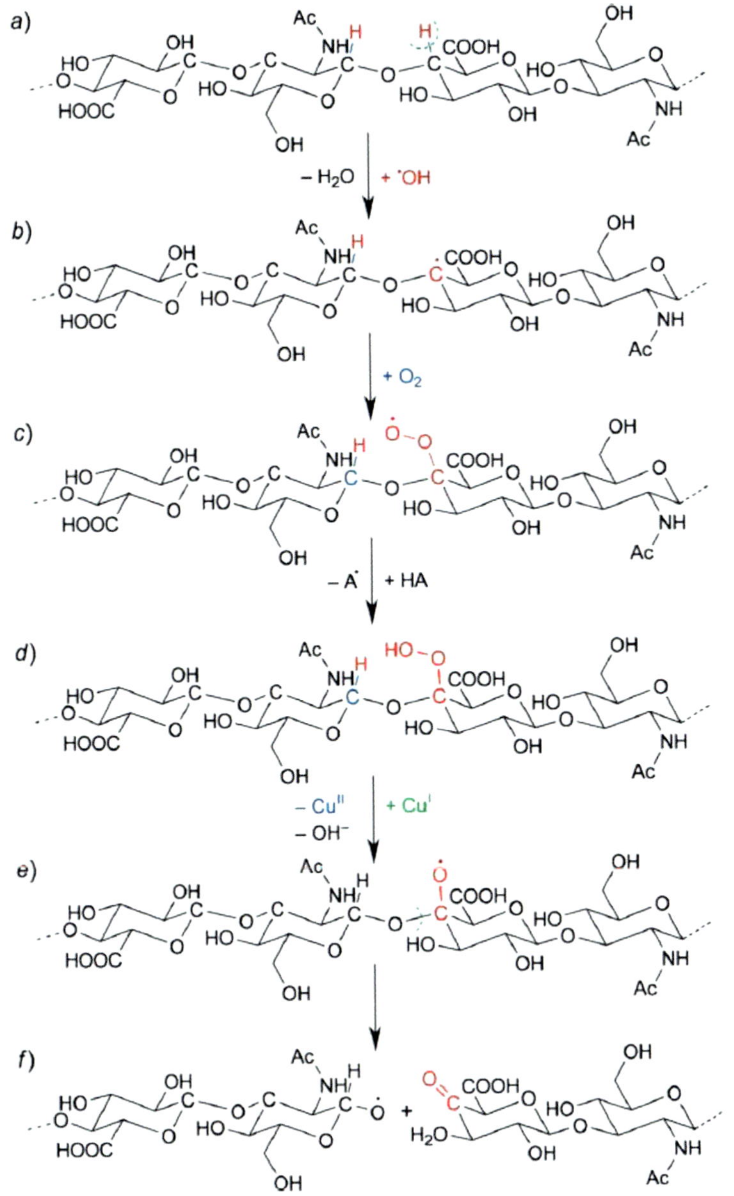

Manganese Porphyrins as Pro-Oxidants

SCHEME 2 Reaction of initiation: (a) an intact HA macromolecule reacts with ˙OH radical; (b) formation of an intermediate, i.e., a *C*-centered HA macroradical. Reactions of propagation and of transfer of the free-radical center; (c) formation of a peroxy-type macroradical; (d) and (e) generation of a HA hydroperoxide and a highly unstable alkoxy-type macroradical. Reaction yielding fragments; (f) an alkoxy-type macroradical and a HA-like macromolecule bearing a terminal C=O group. Both fragments are represented by reduced molar mass.

12.3 METHODS

12.3.1 *STUDY OF HYALURONAN (HA) DEGRADATION*

1. Effects of Mn(II) ions and MnPs on the kinetics of degradation of high-molar-mass HA samples in the presence of ascorbate were studied. After the HA solution was stirred for 30 s, a solution of $MnCl_2$ or MnP (50, 110 or 140 µl) was added, followed by the addition of 50 µl of AA (16 mM). Then, the HA solution was again stirred for 30 s.
2. Anti- and pro-oxidative effects of Mn(II) ions and the MnPs on the kinetics of degradation of high-molar-mass HA samples were studied by using the oxidative system comprising ascorbate and Cu(II) ions (WBOS).

    i. The first experimental set was carried out by adding Mn(II) ions or a MnP at the beginning of the HA treatment: A volume of 50 µl of 160 µM $CuCl_2$ solution was added and stirred for 30 s and left to stand for 7 min 30 s at room temperature (RT). Then 50 µl of Mn(II) ions or of MnP (50–140 µl) were added to the solution and stirred again for 30 s. Finally, 50 µl of AA (16 mM) was added and stirred for 30 s. Then the assayed mixture was immediately loaded into the viscometer cup reservoir. Concentrations of Mn(II) ions and MnPs in the HA solutions in 1 and 2 experimental regimes were 1, 5, 20, and 100 µM.
    ii. The second experimental set involved the same concentrations of the components, each stirred for 30 s, and however, AA was added 1 h later.

12.3.2 *EPR SPECTROSCOPY*

The generation of FRs during HA degradation was examined by spin trapping technique in an EPR X-band EMX spectrometer (Bruker, Rheinstetten, Germany) at ambient temperature. The reaction mixture was composed of HA solution (1.75 mg/ml), Cu(II) ions (1.0 µM), and Mn(II) ions or MnPs (100 µM), andAA(100 µM). The spectra were recorded in the 2nd, 20th, 60th, 90th, or 150th min after the addition of AA.

A 250 µl of each sample solution was thoroughly stirred with 50 µl of 0.212 M DMPO spin trap in H_2O prior to its insert in a thin flat EPR quartz cell. The operational parameters of the equipment were adjusted as follows: center field 3354 G, sweep width 100 G, time constant 81.92 ms, conversion time 20.48 ms, receiver gain 5e+5, microwave power 10 mW, and modulation amplitude 2 G.

12.4 RESULTS AND DISCUSSION

At first, we explored the effect of AA after its addition into the HA solution (Figure 12.1, black curve). We reported a decrease in dynamic viscosity (η) of the HA solution by 1.2 mPa.s (black curve), which was caused by the presence of trace amounts of transition metal ions such as copper and iron in the HA sample. HA absolutely free of transition metals does not exist. A bit more rapid HA degradation was seen after addition of Mn(II) ions in a range between 1 and 100 µM.

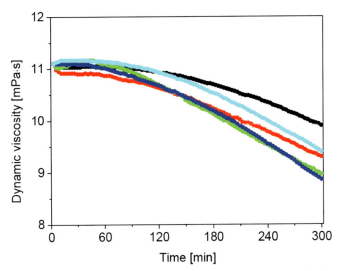

FIGURE 12.1 Time-dependent changes in dynamic viscosity of the HA solution exposed to oxidative degradation by ascorbic acid (100 µM) (black curve) in the presence of Mn(II) ions at concentrations: 1 (red curve), 5 (green curve), 20 (blue curve) and 100 µM (cyan curve).

Next, we subjected HApolymer to oxidative degradation induced by Cu(II) ions (1 µM) and ascorbate (100 µM) (WBOS). Within 5 h-experiment η of the HA solution dropped by 7.71 mPa·s. Then, we examined the effect of Mn(II) ions on HA degradation induced by Cu(II) ions and ascorbate (WBOS) (Figure 12.2). Mn(II) ions at lower concentrations (1 and 5 μM) were potent in part in retarding 'OH radical-induced HA degradation. In this case, Mn(II) ions may most probably function as a displacer of Cu(II) ions. On the other hand, Mn(II) ions at higher concentrations (20 and 100 μM) promoted degradation of HA and the η values were lower by 9.64 and 8.77 mPa·s, respectively. In the second experimental regime (right panel) Mn(II) ions were admixed to the HA reaction mixture 1 h later, where alkoxy- and peroxy-type radicals (Scheme 2) are prevailingly produced. Mn(II) ions (1–20 μM) showed the effect comparable with the reference (black curve). Only at the highest 100 μM concentration (cyan curve), a mild pro-oxidative effect of Mn(II) ions was seen.

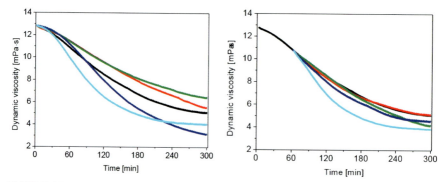

FIGURE 12.2 Time-dependent changes in dynamic viscosity of the HA solutions exposed to oxidative degradation by Cu(II) ions and ascorbic acid (100 μM) (black curve) in the presence of Mn(II) ions at concentrations: 1 (red curve), 5 (green curve), 20 (blue curve) and 100 μM (cyan curve). Mn(II) ions were added before HA degradation began (left panel) or 1 h later (right panel).

Further, we examined the influence of MnP on HA degradation in the presence of ascorbate (100 μM) (Figure 12.3). As evident all three MnP showed dose-dependent pro-oxidative effect. At the highest 100 μM concentration (cyan curves) the degradation of HA was the fastest up to approx. 120 min (A, B) or 60 min (panel C) followed by a remarkable slowing down. The latter observation could be caused by the lack of oxygen in the HA reaction mixture (cf. Scheme 1). As seen, MnTe-2-Pyp5$^+$ (C) was a more potent pro-oxidant than did MnTnBuOE-2-Pyp5$^+$ (A) and MnTnHex-2-Pyp5$^+$ (B). We show that MnPs with ascorbate itself resulted in degradation of HA, thereby we supported the fact that this two-component system produced H_2O_2. We confirmed the presence of H_2O_2 in WBOS-induced HA degradation in the paper by Valachova et al. [25].

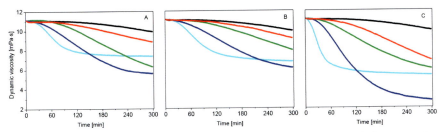

FIGURE 12.3 Time-dependent changes in dynamic viscosity of the HA solution exposed to oxidative degradation by ascorbic acid (100 µM) (black curve) in the presence of MnTnBuOE-2-Pyp5⁺ (A), MnTnHex-2-Pyp5⁺ (B) and MnTe-2-Pyp5⁺ (C) at concentrations: 1 (red curve), 5 (green curve), 20 (blue curve) and 100 µM (cyan curve).

HA was subjected to the action of Cu(II) ions and AA (WBOS), as seen in Figure 12.4, black curve (the reference). Further, we explored the effect of Mn-porphyrin MnTe-2-Pyp5⁺ on oxidatively damaged HA. The addition of this agent at the concentration 1 µM (red curve) led to ˙OH radical-induced HA degradation in a similar manner than in the case of the reference. MnTe-2-Pyp5⁺ at higher levels significantly ameliorated the degradation of HA in a dose-dependent manner up to approx. 120 min (left panel). In other experimental regimes (prevalence of alkoxy- and peroxy-type radicals, right panel) mild degradation was seen in the presence of lower levels of this agent (1 and 5 µM) reaching the decrease in the η values of the HA solutions ca. 4.5 mPa·s. The presence of its higher levels (20 and 100 µM) stimulated HA degradation, whereas the η values of HA solutions were lower by 6.9 mPa·s within 5 h.

Further, the HA solution with WBOS was loaded with MnTnHex-2-Pyp5⁺ (Figure 12.5). MnTnHex-2-Pyp5⁺ at lower concentrations 1 and 5 µM did not affect significantly HA degradation compared to the reference. During 5 h the η value of the HA solutions dropped by 6.53 and 7.66 mPa·s, respectively.

Its addition at higher concentrations (20 and 100 µM) accelerated HA degradation especially up to 120 min, followed by a remarkable slowing down HA degradation (left panel).

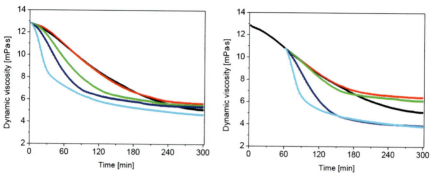

FIGURE 12.4 Time-dependent changes in dynamic viscosity of the HA solution exposed to oxidative degradation by Cu(II) ions (1 μM) and ascorbic acid (100 μM) (black curve) in the presence of MnTe-2-Pyp5$^+$ at concentrations 100 (cyan curve), 20 (blue curve), 5 (green curve) and 1 μM (red curve). MnTe-2-Pyp5$^+$ was added to the reaction mixture before HA degradation begins (left) and 1 h later (right).

The pro-oxidative effect of this Mn-porphyrin was shown also in the system with the alkoxy-/peroxy-type radicals (right panel). Continual decrease in η of HA solutions was seen when using MnTnHex-2-Pyp5$^+$ at lower concentrations 1 and 5 μM. Like in the left panel the degradation of HA molecules significantly promoted in the presence of 20 (blue curve) and 100 μM (cyan curve) of MnTnHex-2-Pyp5$^+$ for about 90 min followed by slow HA degradation (right panel).

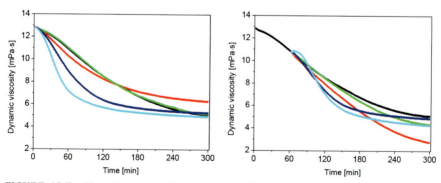

FIGURE 12.5 Time-dependent changes in dynamic viscosity of the hyaluronan solution exposed to oxidative degradation by Cu(II) ions (1 μM) and ascorbic acid (100 μM) (black curve) in the presence of MnTnHex-2-Pyp5$^+$ at concentrations 100 (cyan curve), 20 (blue curve), 5 (green curve) and 1 μM (red curve). MnTnHex-2-Pyp5$^+$ was added to the reaction mixture before HA degradation begins (left) and 1 h later (right).

Manganese Porphyrins as Pro-Oxidants 235

However, a different effect was shown when MnTnBuOE-2-Pyp5$^+$ was examined (Figure 12.6). Unlike two previous MnPs, the addition of MnTnBuOE-2-Pyp5$^+$ accelerated •OH radical-induced HA degradation significantly more rapidly at its lower concentrations (1 and 5 μM) (left panel). On the other hand, no such effect of this porphyrin was shown during alkoxy-/peroxy-type radical-induced HA degradation. In this experimental regime, this MnP showed a moderate pro-oxidative effect in dose-dependent manner (right panel).

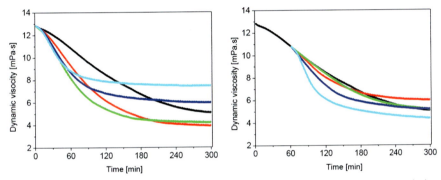

FIGURE 12.6 Time-dependent changes in dynamic viscosity of the hyaluronan solution exposed to oxidative degradation by Cu(II) ions (1 μM) and ascorbic acid (100 μM) (black curve) in the presence of MnTnBuOE-2-Pyp5$^+$ at concentrations 100 (cyan curve), 20 (blue curve), 5 (green curve) and 1 μM (red curve). MnTnBuOE-2-Pyp5$^+$ was added to the reaction mixture before hyaluronan degradation begins (left) and 1 h later (right).

We expressed the percentage of HA degradation induced by WBOS and after the addition of Mn(II) and MnPs (Figure 12.7). Within 60 min HA exposed to WBOS degraded by 15%. A higher rate of HA degradation was reported while examining Mn(II) ions (23.8%), followed by MnTnBuOE-2-Pyp5$^+$ (31.7%), MnTe-2-Pyp5$^+$ and MnTnHex-2-Pyp5$^+$ (44.4%).

We completed our results obtained from rotational viscometry by electron paramagnetic resonance (EPR), which allows detection of adducts of •DMPO-OH (Figure 12.8).

We detected the formation of FRs in the HA solution exposed to Cu(II) ions (1.0 μM), Mn(II) ions (100 μM), and AA (100 μM). Until 90 min we detected the production of ascorbyl radical, which was gradually decreasing and •DMPO-OH adduct began to be visible in the 150th min (A).

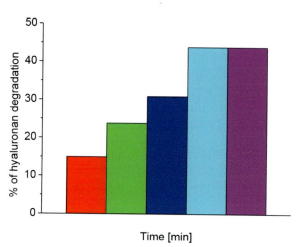

FIGURE 12.7 Percentage of WBOS-induced HA degradation (red) in the presence of Mn(II) ions (green), MnTnBuOE-2-Pyp5+ (blue), MnTe-2-Pyp5+ (cyan) and MnTnHex-2-Pyp5+ (purple) at 100 µM concentration in time 60 min.

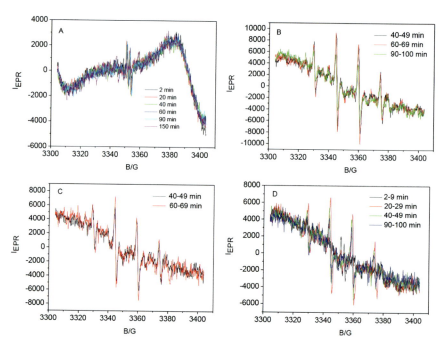

FIGURE 12.8 Time dependence of EPR intensity of ·DMPO-OH adduct observed for the system containing HA, Cu(II) ions (1.0 µM), and ascorbic acid (100 µM) and Mn(II) ions (100 µM) (A), MnTnBuOE-2-PyP^{5+} (B), MnTnHex-2-PyP^{5+} (C) or MnTE-2-PyP^{5+} (D).

Manganese Porphyrins as Pro-Oxidants

In the same HA solution, however, in the absence of Cu(II) ions no radicals were detected (not shown).

As seen in the reaction system composed of HA, Cu(II) ions, MnTnBuOE-2-PyP^{5+}, and AA we confirmed the formation of 'DMPO-OH adduct, whose intensity was the highest during ca. 60 min (B).

Within 60–69 min in the reaction system composed of HA, Cu(II) ions, MnTnHex-2-PyP^{5+}, and AA(C) there was seen a slight production of ascorbyl radical, the maximum amount of 'DMPO-OH adduct was observed after 60 min of reaction, followed by its mild decrease.

The formation of 'DMPO-OH adduct in the reaction system HA, Cu(II) ions, MnTE-2-PyP^{5+}and AA was measured for 100 min, whereas its maximum amount was seen at ca. 30 min (D).

A similar phenomenon, which means pronouncement of HA degradation by the above mentioned Mn-porhyrins was reported by investigating properties of cysteine and *N*-acetylcysteine (NAC) [26, 27]. In the latter case, the observed effect resulted most plausibly from reductive properties of NAC. As EPR measurement showed, NAC in the reaction mixture reduced Cu^{2+} to Cu^+ [27].

The mechanism of MnPs as HA degradation pronouncers, validated by EPR spectrometry (to be published), should be described by following chemical reactions:

$$Mn(III)\text{-porphyrin} + Asc^- \rightarrow Mn(II)\text{-porphyrin} + \text{oxidized ascorbate}$$

$$Mn(II)\text{-porphyrin} + O_2 \rightarrow Mn(III)\text{-porphyrin} + O_2^{\cdot -}$$

$$O_2^{\cdot -} + H_2O_2 \rightarrow O_2 + {}^{\cdot}OH + OH^- \text{ (metal ion catalyzed reaction)}$$

where, the H_2O_2 molecules resulted from Cu(II) catalyzed ascorbate oxidation (Scheme 1).

There are numerous papers pointing to the ability of MnPs to have potential therapeutic applications in cancer [28–30]. Several authors documented effects of the intravenous administration of ascorbate at millimolar levels in treatment of cancer [31–33]. However, such levels of ascorbate have adverse effects on kidney function; therefore, there is an attempt to lower ascorbate dose. One possibility could be to replace ascorbate with MnPs. If only MnPs are present, the anticancer effect is negligible as it is orchestrated through MnP/ascorbate driven H_2O_2 production which in turn oxidizes thiols of signaling proteins. Without ascorbate, there is essentially no effect of ascorbate unless something else produces peroxide like radiation or chemotherapy.

The aim of this study was to replace a transition metal copper in our biogenic system with another transition metal ion. The replacement of Cu(II) with iron ions, such as Fe(III) or Fe(II) in our experiments, caused problems since Cu(III) after reduction to Cu(II) donates $1e^-$, however, complexes Fe(II)/Fe(III) with ascorbate when examining $FeCl_3$ are chemically unstable. A transition metal in a form of $MnCl_3$ applied in WBOS results in the same problem than with $FeCl_3$.

To date in literature, one of the most effective oxidative systems is supposed to be ascorbate and Cu(II) ions under aerobic conditions. Based on our results with Mn-porphyrins we assume that the experiments with Mn(III) ions complexed by porphyrins (e.g., MnTe-2-Pyp5$^+$) could play a role as a strong oxidant, whereas it is evident that toxicity of $CuCl_2$ is a negative factor, while Mn(III)P could be interesting oxidant both *in-vitro* and *in-vivo*. Cancer cells are covered with a layer of high molar mass (HMM) HA, which prevents cells from recognition by components of the immune system. Fragmentation of HMM HA to lower-molar-sized components enhances the probability to recognize cancer cells by immunity cells. Since ascorbate is endogenous (up to 150 µM) we can suggest that WBOS can mimic *in-vivo* conditions. However, whether MnP promotes degradation of HMM HA is it possible to apply such an active principle as a potent anticancer drug? The expectation is that MnP along with ascorbate could significantly decrease molar mass of HA, thereby to remove "the coat of invisibility in cancer cells" [34].

12.5 CONCLUSION

Either Mn(II) ions or MnPs, in the presence of AA, have pro-oxidative effect on HA degradation. If in addition, the Cu(II) ions are present, the HA degradation was further enhanced. All three MnPs have similar redox properties but differ with regards to electrostatic/steric effects. MnTE-2-PyP^{5+} was shown to be the most powerful pro-oxidant compared to other two MnPs due to its smaller size and thus less steric hindrance imposed by longer hexyl and butoxyethyl substitutents. In the presence of Mn(II) tetraaqua ions *vs.* MnPs the degradation of HA proceeds less rapidly at the beginning of the experiment. By performing EPR we detected production of FRs such as ascorbyl radical and 'OH-DMPO adduct.

ACKNOWLEDGMENT

The study was supported by VEGA grant 2/0019/19 and APVV-15-0053 and we thank Dr. Raniero Mendichi from CNR Milano, Italy for determining the molar mass of hyaluronan.

KEYWORDS

- electron paramagnetic resonance
- glycosaminoglycans
- manganese porphyrins
- reactive oxygen species
- rotational viscometry
- transition metals

REFERENCES

1. Vayenas, D. V., Repanti, M., Vassilopoulos, A., & Papanastasiou, D. A., (1998). Influence of iron overload on manganese, zinc, and copper concentration in rat tissues *in-vivo*: Study of liver, spleen, and brain. *Int. J. Clin. Lab. Res., 28*, 183–186.
2. Arredondo, M., Martinez, R., Nunez, M. T., Ruz, M., & Olivares, M., (2006). Inhibition of iron and copper uptake by iron, copper, and zinc. *Biol. Res., 39*, 95–102.
3. Porcheron, G., Garénaux, A., Proulx, J., Sabri, M., & Dozois, C. M., (2013). Iron, copper, zinc, and manganese transport and regulation in pathogenic Enterobacteria: Correlations between strains, site of infection and the relative importance of the different metal transport systems for virulence. *Front Cell Infect. Microbiol., 3*, 90.
4. Weiss, G., & Carver, P. L., (2018). Role of divalent metals in infectious disease susceptibility and outcome. *Clin. Microbiol. Infection, 24*, 16–23.
5. Martinez-Finley, E. J., Gavine, C. E., Aschner, M., & Gunter, T. E., (2013). Manganese neurotoxicity and the role of reactive oxygen species. *Free Radic. Biol. Med., 62*, 65–75.
6. Emsley, J., (2013). In your element: Manganese the protector. *Nature Chem., 5*, 978.
7. Nielsen, F. H., (1999). Ultratrace minerals. In: Shils, M., Olson, J. A., Shike, M., & Ross, A. C., (eds.), *Modern Nutrition in Health and Disease* (9[th]edn., pp. 283–303). Baltimore: Williams & Wilkins.
8. Li, L., & Yang, X., (2018). The essential element manganese, oxidative stress, and metabolic diseases: Links and interactions. *Oxidative Medicine and Cellular Longevity* (p. 11). Article ID 7580707.
9. Skalnaya, M. G., & Skalny, A. V., (2018). *Manganese: Essential Trace Elements in Human Health: A Physician's View* (pp. 107–121). Tomsk: Publishing House of Tomsk State University.

10. Muszynska, A., Palka, J., & Gorodkiewicz, E., (2000). The mechanism of daunorubicin-induced inhibition of prolidase activity in human skin fibroblasts and its implication to impaired collagen biosynthesis. *Exp. Toxicol. Pathol., 52*(2), 149–155.

11. Shetlar, M. R., & Shetlar, C. L., (1994). The role of manganese in wound healing. In: Klimis-Tavantzis, D. L., (ed.), *Manganese in Health and Disease* (pp. 145–157). Boca Raton: CRC Press, Inc.

12. Keen, C. L., Ensunsa, J. L., Watson, M. H., Baly, D. L., Donovan, S. M., Monaco, M. H., & Clegg, M. S., (1999). Nutritional aspects of manganese from experimental studies. *Neurotoxicology, 20*(2/3), 213–223.

13. Ferrer-Sueta, G., Vitturi, D., Batinic-Haberle, I., Fridovich, I., Goldstein, S., Czapski, G., & Radi, R., (2003). Reactions of manganese porphyrins with peroxynitrite and carbonate radical anion. *J. Biol. Chem., 278*(30), 27432–27438.

14. Dua, J., Cullen, J. J., & Buettner, G. R., (2012). Ascorbic acid: Chemistry, biology and the treatment of cancer. *Biochim. Biophys. Acta., 1826*(2), 443–444.

15. Batinic-Haberle, I., Spasojevic, I., Tse, H. M., Tovmasyan, A., Rajic, Z., St. Clair, D. K., Vujaskovic, Z., et al., (2012). Design of Mn porphyrins for treating oxidative stress injuries and their redox-based regulation of cellular transcriptional activities. *Amino Acids, 42*(1), 95–113.

16. Weitzel, D. H., Tovmasyan, A., Ashcraft, K. A., Rajic, Z., Weitner, T., Liu, C., Li, W., Buckley, A. F., et al., (2015). Radioprotection of the brain white matter by Mn(III) N-butoxyethylpyridylporphyrin-based superoxide dismutase mimic MnTnBuOE-2-PyP^{5+}. *Mol. Cancer Ther., 14*(1), 70–79.

17. Ferrer-Sueta, G., Hannibal, L., Batinic-Haberle, I., & Radi, R., (2006). Reduction of manganese porphyrins by flavoenzymes and submitochondrial particles: A catalytic cycle for the reduction of peroxynitrite. *Free Rad. Biol. Med., 41*, 503–512.

18. Shin, S. W., Choi, C., Lee, G. H., Son, A., Kim, S. H., Park, H. C., Batinic-Haberle, I., & Park, W., (2017). Mechanism of the antitumor and radiosensitizing effects of a manganese porphyrin, MnHex-2-PyP. *Antioxid. Redox. Signal, 27*, 1067–1082.

19. Evans, M. K., Tovmasyan, A., Batinic-Haberle, I., & Devi, G., (2013). Manganese porphyrins in combination with ascorbate act as pro-oxidants and mediate caspase-independent cancer cell death. *Cancer Res., 73*(8), 5552.

20. Presti, D., & Scott, J. E., (1994). Hyaluronan-mediated protective effect against cell damage caused by enzymatically produced hydroxyl (OH) radicals is dependent on hyaluronan molecular mass. *Cell Biochem. Funct., 12*, 281–288.

21. Wu, W., Jiang, H., Guo, X., Wang, Y., Ying, S., Feng, L., Li, T., et al., (2017). The protective role of hyaluronic acid in Cr(VI)-induced oxidative damage in corneal epithelial cells. *J. Ophthalmol.,* 6. Article ID 3678586.

22. Chen, H., Qin, J., & Hu, Y., (2019). Efficient Degradation of high-molecular-weight hyaluronic acid by a combination of ultrasound, hydrogen peroxide, and copper ion. *Molecules, 24*, 617.

23. Henderson, E. B., Grootveld, M., Farrell, A., Smith, E. C., Thompson, P. W., & Blake, D. R., (1991). A pathological role for damaged hyaluronan in synovitis. *Ann. Rheum. Dis., 50*, 196–200.

24. Yamazaki, K., Fukuda, K., Matsukawa, M., Hara, F., Yoshida, K., Akagi, M., Munakata, H., & Hamanishi, C., (2003). Reactive oxygen species depolymerize hyaluronan: Involvement of the hydroxyl radical. *Pathophysiology, 9*(4), 215–220.

Manganese Porphyrins as Pro-Oxidants

25. Valachova, K., Topolska, D., Mendichi, R., Collins, M. N., Sasinková, V., & Soltes, L., (2016). Hydrogen peroxide generation by the Weissberger biogenic oxidative system during hyaluronan degradation. *Carbohydr. Polym., 148*, 189–193.

26. Hrabarova, E., Valachova, K., Juranek, I., & Soltes, L., (2012). Free-radical degradation of high-molar-mass hyaluronan induced by ascorbate plus cupric ions: Evaluation of antioxidative effect of cysteine derived compounds. *Chem. Biodivers., 9*, 309–317.

27. Valachova, K., Rapta, P., Juranek, I., & Soltes, L., (2019). On infusion of high-dose ascorbate in treating cancer: Is it time for *N*-acetylcysteine pretreatment to enhance susceptibility and to lower side effects? *Med. Hypotheses, 122*, 8–9.

28. Makinde, A. Y., Luo-Owen, X., Rizvi, A., Crapo, J. D., Pearlstein, R. D., Slater, J. M., & Gridley, D. S., (2009). Effect of a metalloporphyrin antioxidant (MnTE-2-PyP) on the response of a mouse prostate cancer model to radiation. *Anticancer Res., 29*, 107–118.

29. Gad, S. C., Sullivan, D. W. Jr., Spasojevic, I., Mujer, C. V., Spainhour, C. B., & Crapo, J. D., (2016). Nonclinical safety and toxicokinetics of MnTnBuOE-2-PyP5+ (BMX-001). *Int. J. Toxicol., 35*(4), 438–453.

30. Flórido, A., Saraiva, N., Cerqueira, S., Almeida, N., Parsons, M., Batinic-Haberle, I., Miranda, J. P., et al., (2019). The manganese(III) porphyrin MnTnHex-2-PyP^{5+} modulates intracellular ROS and breast cancer cell migration: Impact on doxorubicin-treated cells. *Redox Biol., 20*, 367–378.

31. Alexander, M. S., Wilkes, J. G., Schroeder, S. R., Buettner, G. R., Wagner, B. A., Du, J., Gibson-Corely, K., et al., (2018). Pharmacological ascorbate reduces radiation-induced normal tissue toxicity and enhances tumor radio sensitization in pancreatic cancer. *Cancer Res., 78*(24), 6838–6851.

32. Cieslak, J. A., & Cullen, J. J., (2015). Treatment of pancreatic cancer with pharmacological ascorbate. *Curr. Pharm. Biotechnol., 16*(9), 759–770.

33. Polireddy, K., Dong, R., Reed, G., Yu, J., Chen, P., Williamson, S., Violet, P. C., et al., (2017). High dose parenteral ascorbate inhibited pancreatic cancer growth and metastasis: Mechanisms and a phase I/IIa study. *Sci. Rep., 7*, 17188.

34. Juranek, I., Stern, R., & Soltes, L., (2014). Hyaluronan peroxidation is required for normal synovial function: An hypothesis. Med. Hypothes., 82, 662–666.

35. Andre, P., & Villain, F., (2017). Free radical scavenging properties of mannitol and its role as a constituent of hyaluronic acid fillers: A literature review. *Int. J. Cosmetics Sci., 39*(4), 355–360.

36. Batinic-Haberle, I., & Tome, M. E. (2019). Thiol regulation by Mn porphyrins, commonly known as SOD mimics. *Redox Biology.* doi: https://doi.org/10.1016/j.redox.2019.101139.

37. Hrabarova, E., (2012). Free-radical degradation of high-molar-mass hyaluronan by oxygen free radicals. *Evaluation of Antioxidant Properties of Endogenic Andexogenic Compounds with Thiol Groups in Their Structure (PhD Thesis).* In Slovak, Bratislava: Faculty of Chemical and Food Technology.

38. Mouraviev, V., Venkatraman, T. N., Tovmasyan, A., Kimura, M., Tsivian, M., Mouravieva, V., Polascik, T. J., et al., (2012). Mn-porphyrins as novel molecular magnetic resonance imaging contrast agents. *J. Endourol., 26*(11), 1420–1424.

39. Valachova, K., Hrabarova, E., Priesolova, E., Nagy, M., Banasova, M., Juranek, I., & Soltes, L., (2011). Free-radical degradation of high-molecular-weight hyaluronan induced by ascorbate plus cupric ions. Testing of bucillamine and its SA981-metabolite as antioxidants. *J. Pharm. Biomed. Anal., 56*, 664–670.

CHAPTER 13

Modified Magnetic Metal-Carbon Mesoscopic Composites with Bioactive Substances

V. I. KODOLOV,[1] I. N. SHABANOVA,[2] V. V. KODOLOVA-CHUKHONZEVA,[1] N. S. TEREBOVA,[3] YU. V. PERSHIN,[4] R. V. MUSTAKIMOV,[4] and N. M. POGUDINA[5]

[1]*Scientific Head of BRHE Center of Chemical Physics and Mesoscopics, Professor, M.T. Kalashnikov Izhevsk State Technical University, University in Izhevsk, Russia*

[2]*Professor, Udmurt Federal Research Center, Izhevsk, Russia*

[3]*Doctor, Udmurt Federal Research Center, Izhevsk, Russia*

[4]*Postgraduate, BRHE Center of Chemical Physics and Mesoscopics, Izhevsk, Russia*

[5]*Researcher, BRHE Center of Chemical Physics and Mesoscopics, Izhevsk, Russia*

ABSTRACT

The chapter is dedicated to the consideration of obtaining processes and properties of magnetic copper carbon mesoscopic composites (Cu-C MC), which contains phosphorus and then are modified with the addition of therapeutically active substances such as adenosine tri-phosphoric acid, ascorbic acid (AA), and Urotropine. The producing of magnetic mesocomposites with additive bioactive substances is carried out on the following scheme: (1) the obtaining of copper carbon mesocomposite by mechanic chemical method owing to the joint grating of copper oxide powder with polyvinyl alcohol; (2) the modification of copper carbon mesocomposite by an analogous method with the ammonium polyphosphate (APP) using for the obtaining of phosphorus-containing copper carbon mesocomposite; (3)

the connection of bioactive substances to the phosphorus-containing meso-composite obtained. Phosphorus-containing copper carbon mesocomposite is a magnetic mesoparticle with a phosphorus oxide link for therapeutically active substances. Initial copper carbon mesocomposite consists of copper-containing clusters size of which is equaled to less than 25 nm and which is found in the carbon shell. The above shell contains the carbon fibers from polyacetylene and carbine fragments with unpaired electrons on joints of fragments. At the mechanic chemical interaction of copper carbon meso-composite (MC) with APp, the Phosphorus oxidation state change as well as the change of magnetic characteristics for copper cluster are observed.

The maximum atomic magnetic moment of Copper is equaled to 4.5 μ_B and its obtained at the relation MC to APp correspondent to 1:0.5. In this case, the quantity of unpaired electrons on the carbon shell of nanosized granules is increased in ten times. This fact can testify about the possibility of free radical activity decrease in the striking parts of organism. The connection of above said bio active substances takes place owing to the active carbon shell by means of mechanic chemical method with the small energetic expenses. At the same time, the high magnetic characteristics are preserved at the unpaired electrons presence on the carbon shell. Therefore, the obtained magnetic mesoscopic composites with therapeutically active substances can be considered as transport for bioactive substances and also as the inhibitors of radical processes in organisms.

13.1 INTRODUCTION

At last time, the great attention is spared to the creation of medicine remedies with address direction for the action on the organs which are needed in corre-spondent healing. Usually the transport of therapeutically active substances to the correspondent organ is carried out by means of the magnetic mesoparticles containing the linker connected with medicine substance. In other words, the remedies of medicine carriage in organisms are constructed on the following scheme: transport magnetic mesoparticle-linker-medicine [1–10]. In the above patents the iron-containing nanosized particles, for example, Fe_3O_4, are used as magnetic mesoparticle, and the organic substances connected with this mesoparticle by covalent or coordinative bonds as linker. In this case, it's possible the certain difficulties on the undoing of medicine because of the bioactive substances big interaction with some functional groups. The medicine release from magnetic mesoparticles with linker occurs at the variable magnetic fields (MFs). The best linker can be phosphorus organic

Modified Magnetic Metal-Carbon Mesoscopic Composites

compounds which, as it's known, are easily destructed in water media. Therefore, it's proposed [12, 13] to accomplish the modification of metal containing magnetic mesoparticle by ammonium polyphosphate (APP). At the same time the copper carbon, mesocomposite [14–17] is proposed as magnetic mesoparticle since the Copper has bactericides' and anti-microbes' properties and increases the organism protective forces. For that reason, the copper carbon mesoscopic composite (Cu-C MC) modified by APp. [12, 13] chooses as the investigations object. The modification process is carried out by mechanic chemical method in processing which the Phosphorus reduction and the formation of linker which consists phosphorus with the following oxidation states (OSs) as 0, +3, +5 takes place.

The investigations of phosphorus-containing copper carbon mesocomposite and its analogous are realized with the application of methods complex from which the basic methods are x-ray photoelectron spectroscopy (RPES), transition electron microscopy of high solution (permeation) (TEM), electron microdiffraction, electron paramagnetic resonance (EPR), x-ray diffraction measuring.

According to these investigations, the copper atomic magnetic moment growth is established at the copper carbon mesocomposite modification by APp. and therapeutically active substances. The atomic magnetic moment of Copper is obtained more ($\mu = 4.5\mu_b$) in the comparison with iron-containing nanostructures ($\mu = 2.5\mu_b$). Hence, the proposed magnetic mesoparticles can be interested as the remedies of medicine carriage in organisms by means of drive MF.

The aim of this paper is the consideration of obtaining process of phosphorus-containing Cu-C MCs with connected therapeutically active substances, for example, such, as adenosine tri-phosphoric acid, ascorbic acid (AA) and Urotropine.

13.2 THE COMPOSITION FOR THE TRANSPORT OF MEDICAL ACTIVE SUBSTANCES OF THE NEED ORGANS IN THE ORGANISM

The production of magnetic mesoscopic particles with connected therapeutically active substances is realized by mechanic chemical method on the next scheme:

- The reduction-oxidation synthesis of copper carbon mesocomposite from CuO and polyvinyl alcohol at the reagents relation equaled to 1 mol: 4 mol [11].

- The copper carbon mesocomposite modification by APp. at the regents relation [12, 13].
- The phosphorus containing copper carbon mesocomposite modification by therapeutically active substances, such, as adenosine triphosphoric acid (relation-1:0.02), AA (relation-1:0.2) and Urotropine (relation-1:0.5) (Patent 2018143197).

The mechanic chemical synthesis is carried out with the using of mechanical mortar by the joint grinding of reagents at the energetic expenses approximately equaled to 260–270 kJ/mol. After mechanic, chemical process the mesoscopic product obtained is dried in the closed crucible at 400°C for the first stage, and at 150°C for the second and third stages. Then the product obtained is standee in vacuum at 100–150°C during 3 minutes.

The results of mechanic chemical process with thermo-chemical finishing are estimated with the application of the following methods: RPES, EPR, transition electron microscopy with high permission (TEM HP).

The transition analytic electron microscope FET Tecnai G2F20 with prefix EDAX is used for the investigation of Cu-C MC structure and phase content. High permission corresponds to 2 nm in 1.5 cm (scale).

X-ray photoelectron spectroscopic investigations are realized by x-ray photoelectron magnetic spectrometer with the permission 10^{-4} at the activation AlK_α line 1486 eV in vacuum 10^{-8}–10^{-10} Torre. On the basis of Van Fleck theory, the model for metal atomic magnetic moments calculation is proposed.

EPR investigations are carried out by means of EPR spectrometer E–3 of firm "Varian."

The choice of copper carbon mesocomposite is caused by the copper high atomic magnetic moment because of the formation of unpaired electrons at the reduction oxidation (Red-Ox) process with modifiers. According to the investigations results the Copper atomic magnetic moment growth depends on the number unpaired electrons which participate in Red-Ox process.

The analysis of investigation results with the application of TEM HP shows that the copper carbon mesocomposite obtained represents as mesoscopic particle which consists from metal nucleus (size: 10–20 nm) associated with carbon shell from amorphous carbon fibers (length: 250–1000 nm) (Figure 13.1a).

X-ray photoelectron spectroscopic investigations accompanied with EPR studies show that the carbon fibers contain polyacetylene and carbine fragments with the unpaired electrons on the joints of these fragments (Figure 13.1b, c).

Modified Magnetic Metal-Carbon Mesoscopic Composites 247

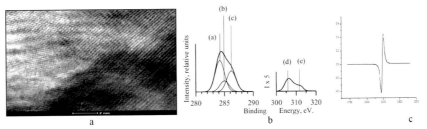

FIGURE 13.1 The TEM image (a), C1s spectrum (b), and EPR spectrum (c) of copper carbon mesoscopic composite.

Copper atomic magnetic moment in above mesocomposite determined on Cu3s spectrum is equaled to 1.3 μ_B.

13.3 THE POSSIBILITIES OF MESOSCOPIC COMPOSITES MAGNETIC PROPERTIES REGULARITY BY MEANS OF THE REDUCTION-OXIDATION REACTIONS

The formation of phosphorus-containing fragments connected with Cu-C MC stimulates the increasing of copper atomic magnetic moment and the creation of linker for the binding with bioactive substances. Some results of Cu-C MC mechanic chemical modification by APp. are shown (Figure 13.2a–c).

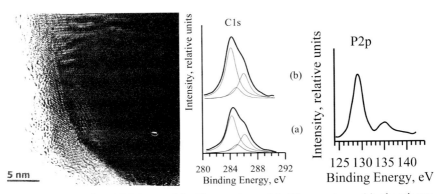

FIGURE 13.2 The TEM image (a), C1s spectrum (b), P2p spectrum (c) phosphorus-containing Cu-C MC.

The binding energy for P2p is changed from 135 eV, correspondent to PO_4 group, to 129 eV for P^0 (Figure 13.2c). It's possible to assume that the

PO$_4$ group reduction process is intensive with the interaction of Copper and Phosphorus. This fact is confirmed by TEM image: the distance between carbon fibers is decreased and some fibers become thick (Figure 13.2a).

C1s spectrum of P modified mesocomposite (Figure 13.2b) is differed from the analogs spectrum of initial mesocomposite on relative intensity C-H line and on the relation of sp^2 and sp^3 components. The intensity of C-H component in C1s spectrum of modified mesocomposite on 15% low in comparison with correspondent intensity in the initial composite spectrum.

According to calculations on Cu3s spectra, the most growth of copper atomic magnetic moment in Cu-C MC modified APp. is arrived at the relation Cu-C MC/APp. equaled to 2. In this case, the copper atomic magnetic moment is equaled to 4.5 μ_B.

At the reagents relation equaled to 1 the copper atomic magnetic moment is decreased more than two times. These results can be explained by the decreasing of electron quantization rate at the APp. layer thickness increasing. The copper atomic magnetic moments values in copper carbon mesocomposite modified by APp. after mechanic chemical modification in the mechanical mortar Retch RM 20 are adduced in Table 13.1.

TABLE 13.1 Copper Atomic Magnetic Moment Values in Systems "Cu-C MC-APP" at the Relations 1:1 and 1:0.5

System Cu-C MC-APP	μ_{cu}
Cu-C MC-APP, relation 1:1	2.0
Cu-C MC-APP, relation 1:0.5	4.5

The experimental studies of copper carbon mesocomposite and system "Cu-C MC-APP" by EPR show the presence of unpaired electrons on the carbon shell. It's noted that the carbon shell of copper carbon mesocomposite modified by APp contains unpaired electrons more than the analogous shell of initial mesocomposite (at the relation of reagents equaled to 1) (Table 13.2). The increase of this relationship leads to the growth of unpaired electrons quantity more than a hundred times. This result corresponds with the growth of magnetic moments (Table 13.1).

The copper atomic magnetic moment growth accompanied with the unpaired electron quantity increasing is explained by the red-ox processes which are carried out with the realization of annihilation and interference phenomena. The results of these processes are illustrated by the comparison of P2p spectra for modified mesocomposites obtained at the relation 1 (1:1) and 2 (1:0.5). The process of phosphorus reduction at the relation equaled to

2 is realized on 50% more [12] in comparison with correspondent process at the relation equaled to 1 (Figure 13.3).

TABLE 13.2 Results of EPR Studies for Copper Carbon Mesocomposite and the Same Mesocomposite Modified by Ammonium Polyphosphate (Relation at Modification Corresponds to 1:1)

Substance	g-Factor	Number of Unpaired Electrons, spin/g
Copper carbon mesocomposite	2, 0036	1.2×10^{17}
System "Cu-C MC-APP"	2, 0036	2.8×10^{18}

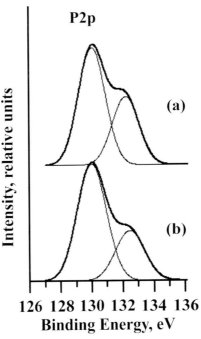

FIGURE 13.3 P2p spectra: (a) Cu-C-APp. (1:1); (b) Cu-C-APp. (1:0.5) [12].

The peak correspondent to binding energy 132.5 eV is ascribed [18] bond C=P that can be possible at the appearance of interference phenomenon because of the direct electromagnetic (EM) field which arises at annihilation in red-ox process. It's possible that the reduced phosphorus is found between the carbon fibers. On this reason, the energy of EM radiation as the result of annihilation stimulates the interference and the covalent chemical bonds formation with the partial conservation of unpaired electrons. The unpaired electrons presence on the mesocomposite carbon shell can promote the decreasing of free radical activity in the defeat part of organism.

It's necessary to note that the conditions change, for instance, the phosphorus-containing fragment quantity or the nature of connected substances leads to the magnetic properties change for the bioactive substances remedies carriage in organisms. The high atomic magnetic moment of Copper (more than $4\mu_B$) can be used for the address carriage of medicine.

13.4 EXAMPLES OF MAGNETIC METAL-CARBON MESOSCOPIC COMPOSITES WITH BIOACTIVE SUBSTANCES

The transport remedy included the copper carbon mesocomposite and linker containing phosphorus for the connection therapeutically active substances is proposed. As examples of further modification the modification processes by such substances as adenosine tri-phosphorus acid (ATP), AA, Urotropine are given.

The investigations of medicines modified by above magnetic mesocomposites are carried out with the application of RPES. Below C1s, spectra for these medicines are presented (Figure 13.4).

The discussion of the electron structures changes for the modified bioactive substances takes place below in the following order: (1) ATP; (2) AA; (3) U.

After the modification of phosphorus-containing copper carbon mesocomposite, obtained at the relation Cu-C MC/APp. equaled to 2, by ATP acid (relation 1:0.02) the following changes of the mesoparticle surface structure takes place:

- In C1s spectrum the maximum for C-H bond is considerably increased;
- In this spectrum the appearance new content at 283 eV correspondent to sp hybridization, and the maximum for sp^2 (C-C) and also sp^3 (C-C) hybridization are decreased;
- In C1s spectrum new maximum (286.7 eV) for C-N bond is observed;
- In Cu2p and P2p spectra the peaks for Cu-P and Cu-P-O are found.

It's necessary to note that the magnetic moment determined on Cu3s spectrum did not exchange and is equaled to 4.5 μ_B.

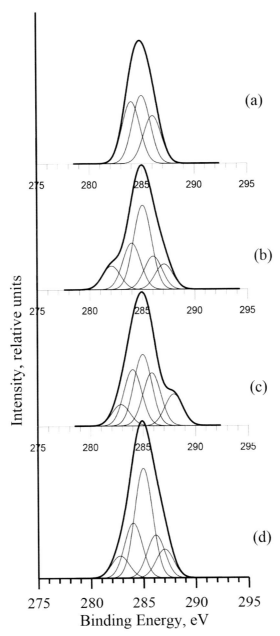

FIGURE 13.4 (a) C1s spectra phosphorus containing copper carbon mesocomposite, (b) and its analogous modified adenosine tri-phosphorus acid (relation 1:0.02), (c) ascorbic acid (relation 1:0.2), (d) Urotropine (relation 1:0.5).

The above changes in spectra can be explained by the carbon phosphorus shell deformation at the interaction of ATP acid with transport remedy.

In the x-ray photoelectron spectra of mesoparticles obtained by means of the phosphorus-containing mesocomposite modification by AA at the reagents relation 1:0.2 the following changes are discovered:

- In C1s spectrum in comparison on the spectrum of initial mesocomposite the new maximum (E_b = 288 eV) for C-OH group is present;
- In Cu2p and P2p spectra the maximums for Cu-P, Cu-O, and P-O bonds are observed.

The magnetic moment determined on Cu3s spectrum also as in above example is not changed and equaled to $4.5\mu_B$.

The changes in x-ray photoelectron spectra of phosphorus-containing Cu-C mesocomposite modified by Urotropine at the relation 1:0.5 also take place. Below the following peculiarities in spectra are given:

- In C1s spectrum the C-H maximum increasing is found;
- In the same spectrum the maximum for C-N bond (E_b = 287 eV) is appeared, and the maximums attributed to sp^2 and sp^3 hybridization have near intensity;
- In P2p spectrum the maximums at 135 eV (P^{+5}) and 129 eV (P^0) take place;
- In N1s spectrum the covalent bond C-N is determined.

In this case, the Copper atomic magnetic moment is near to $4.5\mu_B$.

It's interest to note that the carbine component (sp hybridization) is observed at the mesocomposite quantity increasing.

Thus, the above said examples show on the connection of therapeutically active substances with transport remedy (phosphorus-containing copper carbon mesocomposite) at the conservation of magnetic properties. Latter is necessary for the application of the obtained substances in practice.

13.5 CONCLUSION

In the presented paper, the description of magnetic mesoscopic particles connected with therapeutically active substances for the carriage of correspondent medicine to organs which have a need in these substances is considered. The same order of the above carriage remedies production and the operations succession are proposed. The phosphorus-containing copper carbon mesoscopic composites ($MC_{(P)}$) are applied as magnetic mesoparticles. Then the medicines (M) are linked with $MC_{(P)}$ at the relation ($MC_{(P)}$/M) correspondent to 1:(0.02–0.5) by means of mechanic chemical method with the using the mechanical mortar. The $MC_{(P)}$ applied are obtained by the interaction of copper carbon mesocomposite (Cu-C MC) and APp. at the relation 1:0.5 for the obtaining of high atomic magnetic moment of Copper. The therapeutically active substances such as ATP acid (the relation $MC_{(P)}$/M = 1:0.02), AA ($MC_{(P)}$ /M = 1:0.2) and Urotropine ($MC_{(P)}$ /M = 1:0.5) is linked with phosphorus-containing mesocomposite across the phosphorus-containing liker.

KEYWORDS

- **mesoscopic copper carbon composite**
- **modification by ammonium polyphosphate**
- **therapeutically active substances**
- **unpaired electrons**
- **atomic magnetic moment**
- **polyacetylene and carbine fragments**
- **radical processes inhibitors**

REFERENCES

1. Patent DE 100591 A, Alexiou.
2. Patent DE 4428851 A1.
3. Patent EP 0516252 A2.
4. Patent WO 03/026618 A1.
5. Patent WO 98/58673 (INM).
6. Patent DE 102005016873.6.
7. Patent US 2003028071 A1.
8. RU 2007141588/15.
9. RU 2014154195/15.
10. Patent RU 2490027.
11. Shabanova I. N., Kodolov V. I., Terebova N. S., Kodolova-Chukhontseva V. V., Mustakimov R..V., & Isupov N. Yu. Magnetic Copper Carbon Nanocomposites with Phosphorus containing linker and bioactive systems for the transport of Therapeutic Substances within Organism. – RU 2018122001.

12. Kodolov, V. I., Trineeva, V. V., Kopylova, A. A., et al., (2017). Mechanic chemical modification metal carbon nanocomposites. *Chemical Physics and Mesoscopy, 19*(4), 569–580.
13. Mustakimov R. V., (2019). The method of modification of metal carbon nanostructures by Ammonium Polyphosphate. – Pat. RU 2694092.
14. Kopylova, A. A., & Kodolov, V. I., (2014). Investigations of copper carbon nanocomposite interaction with silicon atoms of silica. *Chemical Physics and Mesoscopy, 16*(4), 556–560.
15. Kodolov, V. I., Khokhriakov, N. V., Trineeva, V. V., & Blagodatskikh, I. I., (2010). *Problems of Nanostructure Activity Estimation, Nanostructures Directed Production and Application. Nanomaterials* (pp. 1–18). Yearbook – 2009, from nanostructures, nanomaterials, and nanotechnologies to Nanoindustry. NY. Nova Science Publishers, Inc.
16. Chashkin, M. A., Kodolov, V. I., Zakharov, A. I., et al., (2011). Metal/carbon nanocomposites for epoxy compositions: Quantum-chemical investigations and experimental modeling. *Polymer Research Journal, 5*(1), 5–19.
17. Mustakimov, R. V., Kodolov, V. I., Shabanova, I. N., & Terebova, N. S., (2017). Modification of copper carbon nanocomposites by means of ammonium polyphosphate for the application as modifiers of epoxy resins. *Chemical Physics and Mesoscopy, 19*(1), 50–57.
18. Wang, J. Q., Wu, W. H., & Feng, D. M., (1992). *The Introduction to Electron Spectroscopy (XPS/XAES/UPS)* (p. 640). Beijing: National Defense Industry Press.

CHAPTER 14

Pre-Treatment of Rusted Steel Surfaces with Phosphoric Acid Solutions

VICTOR M. M. LOBO and ARTUR J. M. VALENTE

CQC, Department of Chemistry, University of Coimbra, 3004-535 Coimbra, Portugal, E-mail: vlobo@ci.uc.pt (V. M. M. Lobo)

ABSTRACT

Steel surfaces in large industrial installations are exposed to corrosion phenomena. Corrosion prevention involving painting coats need a good pre-treatment of the metallic surface to guarantee a lasting corrosion protection. Painting coatings are more effective if they acquire strong intermolecular forces with the steel. H_3PO_4 solutions may be used for such pre-treatment giving good results as a consequence of "rust conversion" by H_3PO_4. A deep and extensive physico-chemical research study of the interaction of H_3PO_4 with the multitude of iron oxides, carbonates, sulfides ("rust"), etc., that cover the metallic surface, even after "normal" cleaning, previous to the painting process, is here reported.

The nature of the solutions and the processes of their application recommended in the scientific and technical literature vary significantly. We report here the chemical and structural nature of those corrosion products (rust) formed in a variety of conditions, and their interaction with H_3PO_4 solutions (rust conversion products) at different concentrations. The present studies show that a 14.6% (m/m) H_3PO_4 solution would be adequate.

14.1 INTRODUCTION

Steel surfaces are part of large industrial installations, bridges, liquid, and gas tanks, pipelines, ships, etc., (Figure 14.1). Thus, they are exposed to corrosion phenomena.

FIGURE 14.1 Photographs of large industrial installations.

Corrosion prevention often consists of appropriate painting coats, after an adequate primary coat [1]. However, a good pre-treatment of the metallic surface is very important to guarantee a lasting corrosion protection. Painting coatings are more effective if they acquire strong intermolecular forces with the steel substrate [2].

It has long been empirically known that H_3PO_4 solutions may be used for such pre-treatment giving good results as a consequence of "rust conversion" by H_3PO_4. But, there was not a deep and extensive physico-chemical research study of the interaction of H_3PO_4 with the multitude of iron oxides, carbonates, sulfides ("rust"), etc., that cover the metallic surface, even after "normal" cleaning, previous to the painting process[3, 4].

There was not a clear picture of the chemical and structural composition of this rust, which obviously varies, not only with the process and degree of the cleaning treatment, but also with the time allowed for its formation, and the atmospheric conditions (e.g., industrial, industrial-marine, severe-marine, rural atmospheres).

The nature of the solutions and the processes of their application recommended in the scientific and technical literature vary significantly and sometimes even contradictorily [4–6].

It has been our aim to clarify the chemical and structural nature of those corrosion products (rust) formed in a variety of conditions, and their interaction with H_3PO_4 solutions (rust conversion products) at different concentrations eventually with the addition of $Al(OH)_3$.

Apart from this scientifically fundamental objective, it was our ultimate wish to propose the best treatment conditions, including H_3PO_4 concentration of the solutions, and processes of application.

Pre-Treatment of Rusted Steel Surfaces with Phosphoric Acid Solutions

In aqueous solutions, and in natural circumstances, the corrosion of ferric alloys may be interpreted as the oxidation of iron, symbolically represented by:

$$2\ Fe(s) \rightleftarrows 2\ Fe^{2+}(aq) + 4e^- \tag{14.1}$$

$$O_2(aq) + 4\ H^+(aq) + 4e^- \rightleftarrows 2\ H_2O(l) \tag{14.2}$$

$$2\ Fe(s) + O_2(aq) + 4\ H^+(aq) \rightleftarrows 2\ Fe^{2+}(aq) + 2\ H_2O(l) \tag{14.3}$$

where, Eqs. (14.1) and (14.2) would represent the anodic and cathodic reactions, respectively, and (14.3) their corresponding sum. $Fe^{2+}(aq)$ eventually is oxidized to $Fe^{3+}(aq)$ and considering that there are species derived from the atmosphere (CO_2, SO_2, H_2S, etc.), the molecular composition of rust, that is, the molecular species originated as a consequence of (14.3), is certainly very complex.

The Gibbs energy, ΔG (at T=cte., P=cte.), characterizing the Eq. (14.3) is negative, that is, under natural conditions, (14.3) occurs spontaneously.

Assuming that there is almost always, a film of adsorbed H_2O on the surface of all ferric alloys exposed in the atmosphere, Eqs. (14.1), (14.2), and (14.3) may equally interpret the atmospheric corrosion of iron.

On the other hand, the ionization of phosphoric acid (H_3PO_4), in aqueous solutions may be symbolically expressed by:

$$H_3PO_4(aq) \rightleftarrows H_2PO_4^-\ (aq) + H^+(aq) \tag{14.4}$$

$$K_1 = 7.1 \times 10^{-3}$$

$$H_2PO_4^-\ (aq) \rightleftarrows HPO_4^{2-}\ (aq) + H^+(aq) \tag{14.5}$$

$$K_2 = 6.3 \times 10^{-8}$$

$$HPO_4^{2-}\ (aq) \rightleftarrows PO_4^{3-}\ (aq) + H^+(aq) \tag{14.6}$$

$$K_3 = 4.4 \times 10^{-13}$$

$$H_3PO_4(aq) \rightleftarrows PO_4^{3-}(aq) + 3H^+(aq) \tag{14.7}$$

where, K values represent the corresponding ionization constants. Obviously, the extent of Eqs. (14.4), (14.5), or (14.6) depends on the pH and interactions of the hydrophosphate ions with other species in the solution. The interaction with oxidized iron tends to form complex structures that minimize further corrosion of iron.

Mild steel panels, duly prepared, were exposed in three different types of environment: atmosphere with a relative low degree of contamination; atmosphere with a high content of chloride salts; atmosphere with a high content of SO_2. The nature of the corrosion products, that is, the surface characterization, was studied using powerful analytical tools.

14.2 EXPERIMENTAL

The experimental work was carried out using mild steel panels (150 cm^2 of area) prepared according to ISO 630 (Structural Steels, International Standard Organization, Geneva, Switzerland, 1995) [7]; three different types of environment were chosen in order to mimetize, in as much as possible, the actual conditions where the large anthropogenic metallic surfaces are subjected to natural atmospheric corrosion. The locations were specimens were allowed to be corroded are characterized as follows:

1. Atmosphere with a relative low degree of contamination (Lumiar, N38° 46.227'; W09° 10.769') (Figure 14.2A).
2. Atmosphere with a high content of chloride salts (Sines, N37° 57, 430'; W08° 52, 183') (Figure 14.2B).
3. Atmosphere with a high content of SO_2 (Barreiro, N38° 39, 880'; W09° 04, 499') (Figure 14.2C).

The exposure time of panels (in months) was such that the panels would attain a degree of rusting C (ISO-8501-1). The chosen corrosion periods were 1, 3, 10, and 24 months.

The panels were treated with 7.55%, 14.6%, 21.1%, 27.4%, 33.2%, 38.6%, 43.8%, 53% and 79.8% (w/w) H_3PO_4 solutions in the following conditions: for each sample with a 150 cm^2 area of corroded and subsequently well-brushed steel surface, 0.7 mL of solution were applied on the surface, with a cross-wise motion using a small perfectly clean and dry brush. Drying of the applied pre-treatment was undertaken at $23 \pm 2°C$ and $50 \pm 5\%$ relative humidity over 24 h.

Once taken to the laboratory, they were again brushed using an automatic and reproducible technique to a degree CSt2 (SIS055900, now also in ISO 8501), degreased, in wet cabinet (DIN 50017), for 8 days, and dried at 50°C for 48 hours.

FIGURE 14.2 Photos of the location of panels in different atmospheres.

14.2.1 SURFACE CHARACTERIZATION

Surfaces were observed visually and with an Olympus SHZ optical microscope equipped with an Olympus OM-2SP camera (Figure 14.3). They were also examined with a JEOL JSM-35 CF scanning electron microscope (SEM) attached to an energy-dispersive X-ray spectrometer (Tracor Northern). The samples were previously coated with a carbon film using a JEOL JEE-4X vacuum evaporator.

FIGURE 14.3 Optical microscope equipment and a corresponding representative photo.

XRD was performed directly on the steel surface using a Rigaku D/MAX II C diffractometer (Figure 14.4) operating at 45 kV/20 mA with a curved graphite crystal monochromator and a broad focus cobalt source. The samples were scanned at a speed of 2 min^{-1} over the range 8–85°.

Infrared (IR) absorption spectra of the surface corrosion products were obtained using a Perkin-Elmer model 598 instrument and an FTIR Nicolet 55 DXC spectrometer on KBr pellets.

Mössbauer spectra were obtained with the samples at room temperature (RT) using a ^{57}Co in Rh source. Power samples were enclosed in Perspex holders and measured by transmission geometry, while the panels were measured by back-scattering geometry.

Solubility measurements of the rust conversion layers (RCL): 18 mL of H_2O were left in contact with RCL for 24 hours; the Fe and P content were determined by atomic mass spectrometry (AMS).

The paint adhesion was evaluated according to ISO 4624. Under these conditions, the results were assessed as breaking force (expressed in Newton's), breaking strength (in MegaPascals), and nature of the failure. An Instron instrument series 1100 of 10 Ton, connected to a Hewlett Packard HP 85 microcomputer, was used for the pull-off test.

FIGURE 14.4 Photography of the XRD equipment.

14.3 EXPERIMENTAL RESULTS AND DISCUSSION

With the indicated variables in H_3PO_4 (concentration ca. 7, 14, 21, 27, 33, 38, 43, 53, and 79%, exposure time 1, 3, 10, and 24 months, and 3 different types

Pre-Treatment of Rusted Steel Surfaces with Phosphoric Acid Solutions 261

of atmospheres) we naturally have an amount of information which would be tedious to present here. Therefore, we only present representative cases.

Figures 14.5 and 14.6 show the results of microscopic observation (visual and SEM) of the pretreated surfaces together with the respective EDS spectra (10 months, low contamination).

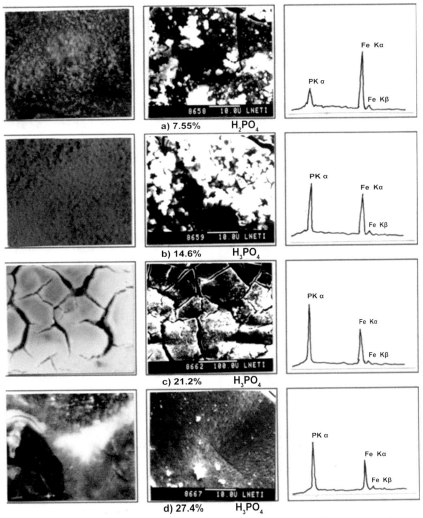

FIGURE 14.5 Some results obtained with well-brushed steel surfaces previously rusted by exposure of 10 months to the atmosphere followed by pretreatment with phosphoric acid solutions of different concentrations (from 7.55 to 27.4% (w/w)). (Left column: micrograph (66×); central column: SEM micrograph; right column: EDS spectra).

Source: Adapted with permission from Ref. [4]. © 1993 Elsevier.

From the analysis of Figures 14.6 and 14.7, we can see that with H_3PO_4 7.55% (m/m) not all the rust formed in 10 months exposure has been converted, and therefore this would not be a sufficient concentration for the purpose of the intended pre-treatment.

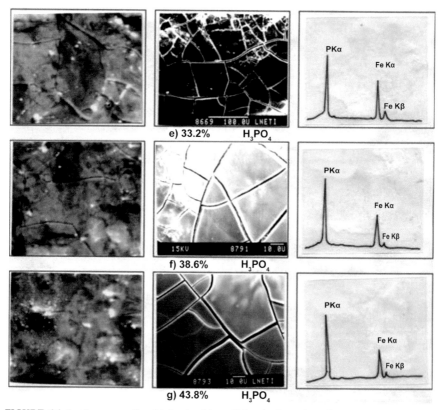

FIGURE 14.6 Some results obtained with well-brushed steel surfaces previously rusted by exposure of 10 months to the atmosphere followed by pretreatment with phosphoric acid solutions of different concentrations (from 33.2 to 43.8% (w/w)).(Left column: micrograph (66×); central column: SEM micrograph; right column: EDS spectra).

Source: Adapted with permission from Ref. [4]. © 1993 Elsevier.

A 14.6% (m/m) H_3PO_4 solution would be adequate; 27.4% would be already higher than desirable, and 33.2% forms vitreous dark-brown films, showing cracks and too high acidity for the present objectives and therefore undesirable. These other results show that higher concentrations are worst, e.g., at 53% we have what Krebs calls phosphate glass [8].

Pre-Treatment of Rusted Steel Surfaces with Phosphoric Acid Solutions 263

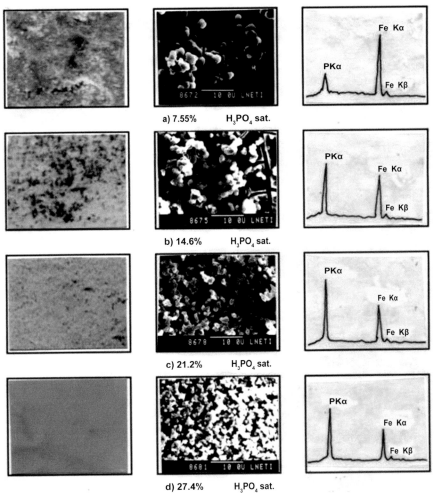

FIGURE 14.7 Some results obtained with well-brushed steel surfaces previously rusted by exposure of 10 months to the atmosphere followed by pretreatment with solutions containing saturated phosphoric acid and aluminum hydroxide of different concentrations (from 7.55 to 27.4% (w/w)).(Left column: micrograph (66×); central column: SEM micrograph; right column: EDS spectra).

Figures 14.7 and 14.8 show that the pretreatment substrate surfaces with H_3PO_4 saturated solution containing $Al(OH)_3$ lead to a significant beneficial effect.

The morphology of the studied RCLs is quite different. The lower acid solution concentration leads to the formation of a white-bluish RCL with a granular structure. The addition of Al(OH)$_3$ to H$_3$PO$_4$ solutions appears to lead to RCLs with a smaller grain size in a more compact structure.

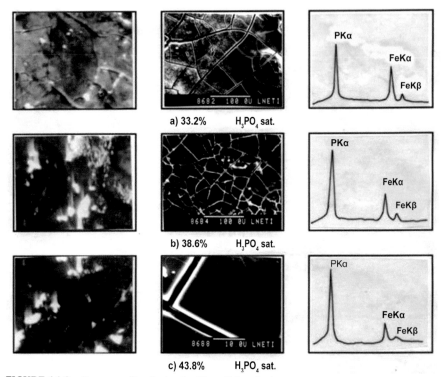

FIGURE 14.8 Some results obtained with well-brushed steel surfaces previously rusted by exposure of 10 months to the atmosphere followed by pretreatment with solutions containing saturated phosphoric acid and aluminum hydroxide of different concentrations (from 33.2 to 43.8% (w/w)).(Left column: micrograph (66×); central column: SEM micrograph; right column: EDS spectra).

The crystalline species identified in these RCLs by XRD (Figure 14.8) are mainly hydroxides-lepidocrocite, γFeOOH, and minor goethite, αFeOOH- and iron phosphates. Occasionally, a spinel-type oxide (γFe$_2$O$_3$/Fe$_3$O$_4$) was also detected. No crystalline phase-specific for aluminum could be traced in the XRD spectra. A full characterization is described in Table 14.1.

Broadbands were noticed in the XRD spectra of the white-bluish RCL, positioned around the main reflection of vivianite, a ferrous phosphate hydrated phase with the formula Fe$_3$(PO$_4$)$_2$.8H$_2$O which was detected by IR spectrometry (not shown) (Figure 14.9).

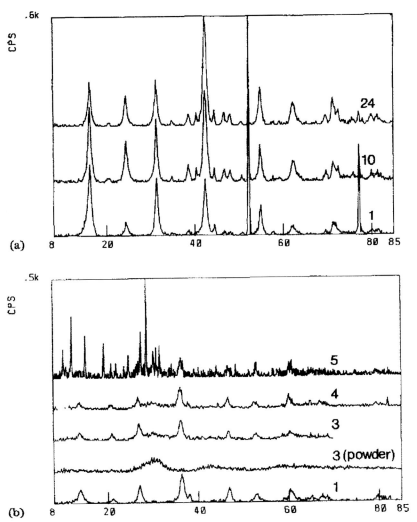

FIGURE 14.9 X-ray diffractograms of (a) surface corroded for different exposure time and subsequently well brushed, and (b) surfaces corroded for 10 months, followed by brushing before and after pretreatment at different H_3PO_4 concentrations.

Source: Adapted with permission from Ref. [4]. © 1993 Elsevier.

On the other hand, higher concentrations of H_3PO_4 (with or without $Al(OH)_3$) lead to the formation of RCLs with a vitreous brown-black color and an XRD pattern indicating an amorphous state. This glassy state is known as "phosphate glass," being attributed to disordering by rotation about hydrogen bridges in the material resulting from concentrated acid solutions.

TABLE 14.1 Characterization of Compounds Occurring in Samples Corroded for 10 Months

	Chemical Formula	Crystallo–graphic System	Characteristics Bands		Mössbauer Parameters		
			EAIV/cm^{-1}	DRX	BT	δ	ΔE_0
Lepidocrocite	γ-FeOOH	orthorhombic	3120–2857(f) 1161–1145(d) 1020–1013(f) 753–738(f) 881(d) 590(d) 520(m) 483–465(m) 360(f)	d=6.26 d=3.29 d=2.47 File 8–98	46	0.47	<0.1
Goethite	α-FeOOH	orthorhombic	890–882(f) 797–793(m) 643–610(m) 599(m) 450(m) 400(f) 372(d)	d=4.18 d=2.45 d=2.69 File 29–713	50	0.6	–
Magnetite	Fe$_3$O$_4$	cubic	580(f, l) 450(f)	d=2.53 d=1.48 d=2.97 File 19–629	51.6	0.42	0.06
Akaganeite	βFeOOH	tetragonal	3450(f, l) 3350(f, l) 1630(d, l) 1520(d, l) 1420(d, l) 1150(md) 970(md) 840(ml) 650(f, ml) 450(f, ml)	d=3.33 d=2.55 d=7.47 File 34–1266	48	0.47	–
Hematite	αFe$_2$O$_3$	hexagonal romboedric	3435(fl) 1620(ml) 575(fl) 485(f) 360(f)	d=2.70 d=2.52 d=1.69 File 33–664	54	–	–
Crystalite	δFeOOH	hexagonal	3420–2850(fl) 3320(m) 1510(mbl) 1125(ml) 915(ml) 800(d) 690–670(m) 660(d) 530(dl) 475(fl) 420(d) 325(ml) 285(dl) 280(dl) 280(dl) 225(mdl)	d=2.55 d=2.26 d=1.69 File 13–87	–	–	–
Iron oxi-hydroxide amorphous	FeOx(OH)$_{3+}$ 2x x=0.4	–	3380(fl) 1460(ml) 1325(ml) 930(mdl) 800(mdl) 710(ml) 585(dl) 465(fl) 420(fl) 280(ml)	–	–	–	–
Maghemite	γFe$_2$O$_3$	cubic	3460(ml) 1620(ml) 1100(ml) 1020(md) 1030(md) 725(h) 690(f) 640(lf) 560(lf) 485(m) 445(m) 420(m) 398(f) 370(m) 330(m) 265(d) 238(d)	d=2.52 d=1.48 d=2.95 File 24–81	52.5	0.45	<0.05

The best behavior of RCLs obtained with phosphoric acid solutions saturated with $Al(OH)_3$ can probably be explained by the solubility properties of the newly-formed layers.

Painting coatings are widely used for corrosion prevention. For best results, a pre-treatment based on H_3PO_4 solutions will contribute to a "rust conversion" process that will strength intermolecular forces between the painting coats and the steel substrate, if adequately performed, mainly with respect to concentration.

The nature of the corrosion products varies significantly with the time of exposure and type of the atmosphere and the nature of the rust conversion products varies considerably with the H_3PO_4 concentration and is improved with the addition of $Al(OH)_3$. Therefore, the H_3PO_4 pre-treatment should have the above parameters in consideration. However, in most common circumstances, a H_3PO_4 concentration from around 15% (m/m) to 20% (m/m) is the most appropriate. The larger the exposure time, the less effective is the converting action of H_3PO_4 due to the growth of goethite. Addition of $Al(OH)_3$ so as to form a saturated solution improves the action of the pre-treatment with H_3PO_4.

KEYWORDS

- **corrosion**
- **phosphoric acid**
- **rust conversion**
- **scanning electron microscope**
- **rust conversion layers**
- **atomic mass spectrometry**

REFERENCES

1. Kendig, M., & Mills, D. J., (2017). An historical perspective on the corrosion protection by paints. *Prog. Org. Coatings., 102*, 53–59. doi: 10.1016/j.porgcoat.2016.04.044.
2. Gao, X., Li, W., & Ma, H., (2017). Effect of anti-corrosive performance, roughness, and chemical composition of pre-treatment layer on the overall performance of the paint system on cold-rolled steel. *Surf. Coatings Technol., 329*, 19–28. doi: 10.1016/j.surfcoat.2017.09.029.

3. Santana, I., Pepe, A., Jimenez-Pique, E., Pellice, S., & Ceré, S., (2013). Silica-based hybrid coatings for corrosion protection of carbon steel. Part I: Effect of pretreatment with phosphoric acid. *Surf. Coatings Technol., 236*, 476–484. doi: 10.1016/j. surfcoat.2012.07.086.

4. Almeida, E., Pereira, D., Waerenborgh, J., Cabral, J. M. P., Figueiredo, M. O., Lobo, V. M. M., et al., (1993). Surface treatment of rusted steel with phosphoric acid solutions: A study using physico-chemical methods. *Prog. Org. Coatings., 21*, 327–338. doi: 10.1016/0033–0655(93)80048-F.

5. Frondistou-Yannas, S., (1986). No title. *J. Prot. Coatings Linings, 3*, 26.

6. Hendry, M., (1985). *J. Prot. Coatings Linings, 2*, 18.

7. ISO 630 (1995). Structural steels – plates, wide flats, bars, sections and profiles. International Standard Organization, ISO, Geneva.

8. Krebs, H., (1968). Fundamentals of Inorganic Crystal Chemistry, McGraw-Hill, New York.

CHAPTER 15

Transport Phenomena of Electrolytes in Aqueous Solutions: Concepts, Approaches, and Techniques

ANA C. F. RIBEIRO,[1] EDUARDA F. G. AZEVEDO,[1]
ANA PAULA COUCEIRO FIGUEIRA,[2] and VICTOR M. M. LOBO[1]

[1]Centro de Química, Department of Chemistry, University of Coimbra, 3004-535 Coimbra, Portugal, E-mail: anacfrib@ci.uc.pt (A. C. F. Ribeiro)

[2]Faculty of Psychology and Educational Sciences, University of Coimbra, 3004-535 Coimbra, Portugal

ABSTRACT

Concepts, approaches, and techniques involved in the determination of the transport properties (TP) in solutions (diffusion, conductivity, and viscosity), their importance in the scientific and technologic communities, and the critical analysis of these data are focused in this work.

15.1 ELECTROLYTES

Different definitions about the concept of electrolyte are found in the literature [1, 2]. For example, P. Radsahakrishnamurty considers that electrolyte is a substance which conducts electric current through the agency of ionic species (and not electrons). However, this definition is polemic, leading Lobo [1] to reach the following comment "*apparently it would not consider, e.g., a crystal sodium chloride or the gas hydrogen chloride as being electrolytes because either of them hardly conducts any current in normal circumstances*," and to state the following definitions, which we have accepted until now. That is;

*"**Electrolyte:** A substance when dissolved in a given solvent produces a solution with an electric conductivity higher than the solvent conductivity. Maybe a solid (e.g., sodium chloride), a liquid (e.g., sulfuric acid), or a gas (e.g., hydrochloric acid).*

Just for short, but meaning the above, we could say:

***Electrolyte:** A substance that increases the electrical conductivity of the solvent.*"

15.2 TWO APPROACHES TO DEFINE TRANSPORT PROPERTIES (TP): PHENOMENOLOGIC, AND USING THERMODYNAMICS OF IRREVERSIBLE PROCESSES (TIP)

15.2.1 *PHENOMENOLOGY OF TRANSPORT PROPERTIES (TP)*

The transport processes in the aqueous solutions result from a situation in which these systems are far from their thermodynamic equilibrium states, caused for example by heterogeneities of temperature, composition, or electrical potential. In each case a flow of a system variable, J, is associated with the corresponding macroscopic quantity gradient, X:

$$J = -\text{K grad } X \tag{15.1}$$

where, K represents the transport coefficient (e.g., mutual diffusion coefficient for mass transport [3–7]) (Table 15.1). Through the relation obtained by Fourier [3–7] and the similarity established initially by Berthollet between diffusion and heat flow [3–7], Fick established that the force responsible for the flow in the process of diffusion in a binary system is a concentration gradient [3–7]. The measurements of transport properties (TP) require laborious attention and particular attention to possible sources of error, so that they can be obtained with the accuracy and precision required by science and technology.

15.2.2 *THERMODYNAMICS OF IRREVERSIBLE PHENOMENA (TIP): DOES IT MAKE SENSE TO APPLY THERMODYNAMICS TO SYSTEMS WHICH ARE NOT IN EQUILIBRIUM?*

Thermodynamics is a remarkably intellectual structure [9] because it deals with mathematical relationships between macroscopic properties which we

Transport Phenomena of Electrolytes in Aqueous Solutions 271

can measure in the laboratory, such as temperature, volume, and solubility, being independent of theoretical models of the microscopic nature of matter. Thermodynamics is one of the most powerful techniques at our disposal for the study of natural phenomena. It concerns to balancing positions.

TABLE 15.1 Examples of Transport Properties

Transport Property	Driving Force	Flow Equation	Coefficient
Mass	Concentration Gradient	$J_m = -D\,(dc/dx)$	Diffusion coefficient, D
Energy/ heat	Temperature gradient	$J_h = -k\,(dT/dx)$	Thermal conductivity, k
Momentum	Velocity gradient	$J_n = -\eta\,(dv/dx)$	Viscosity, η
Electric	Potential gradient	$J_n = -\sigma\,(d\Phi/dx)$	Electrical conductivity, s

A generic treatment of mass transport processes in electrolyte solutions may be done using the thermodynamics of irreversible or non-equilibrium phenomena (TIP). Miller [10–15] considered non-equilibrium thermodynamics a linear extension of classical thermodynamics.

TIP is based on the hypothesis that classical concepts, such as temperature and entropy, are valid in small subvolumes which are assumed to be in "local" equilibrium. The absence of perturbations in the local equilibrium is assumed when temperature, composition, or temperature gradients occur, as well as the transport of heat, matter, or electricity through each sub-volume. These hypotheses have been experimentally tested in many physical and chemical processes, in particular for transport processes in electrolyte solutions. This is a general linear description of all types of transport processes. Specifically, it takes into account the effects of each type of flow on all others. In other words, the application of thermodynamics carried out in the works of Gibbs and later Lewis and Randall deal almost entirely with states of equilibrium, and with reversible transitions between them [10–15]. Such transitions occur in a continuous set of equilibrium states by infinitely slow processes and can be reversed by an infinitely small variation in the variables that determine the nature of the system. Thus, classical thermodynamics can be successfully applied to some systems in which irreversible processes occur. More generally, if irreversible processes occur sufficiently slowly at the boundaries between homogeneous phases so that the mechanical and thermal equilibrium remains within the phases, the multiphase system can be treated thermodynamically. For homogeneous systems, the problem is solved by the local equilibrium hypothesis. The system is assumed to be divided into the identical number of subsystems, macroscopically small but microscopically large, each at constant volume and in internal thermodynamic equilibrium. If this procedure is valid, it is possible to calculate the rate of entropy production in homogeneous and heterogeneous systems (dS_{int}/dt) subject to irreversible processes. In other words, the rate of entropy production can be represented by a sum of products of density of flux J_i and generalized and independent "thermodynamic forces" X_i. This product is referred to as the dissipation function, expressing, for an n-component system, by:

$$\varnothing = T\frac{dS_{int}}{dt} = \Sigma = J_i X_i \tag{15.2}$$

where T is the temperature; J_i represents the fluxes of the ions or the solvent and X_i is the thermodynamic force.

This makes it possible to establish the minimum number of independent variables required to define a non-equilibrium system (provided it is not too far from equilibrium) and to relate these variables to experimentally determined quantities.

Another hypothesis on which the thermodynamics of irreversible processes (TIP) is based is that the flux densities, J_i, can be related to the thermodynamic "forces," X_j, through a set of phenomenological linear equations, which in the case of a isotropic system, have the form being L_{ii} the independent phenomenological coefficients related to the coefficients of thermal or electrical conductivity or diffusion, while those of the form L_{ij} are associated as effect of force X_j on the density flux J_i.

Another hypothesis on which the TIP is based is that the flux densities, J_i, can be related to the thermodynamic "forces," X_j, through a set of phenomenological linear equations, which in the case of an isotropic system, have the form (15.3):

$$J_i = \sum_{j=1}^{n} L_{ij} X_j \tag{15.3}$$

represent the independent phenomenological coefficients related to the coefficients of thermal or electrical conductivity or diffusion, while those of the form are associated as effect of force on the density flux J_i.

This equation is of great importance for systems where there are several forces at the same time, provided a second concept is accepted. That is,

Transport Phenomena of Electrolytes in Aqueous Solutions 273

according to Onsager the number of is subsequently reduced by establishing that the matrix of coefficients || is symmetrical, i.e., $L_{ij} = L_{ji}$.

In conclusion, two approaches are referred to describe the TP. In the case of the isothermal diffusion of electrolytes in aqueous solutions the Fick's laws and thermodynamics applied to irreversible processes (TIP) are used.

15.3 TRANSPORT PROPERTIES (TP): VISCOSITY, CONDUCTIVITY, AND DIFFUSION

15.3.1 APPLICABILITY OF THE TRANSPORT PROPERTIES (TP) TO BIOLOGIC SYSTEMS OF THE INTEREST

TP in aqueous electrolyte solutions have been measured in different conditions (different electrolytes, concentration, temperature, techniques used, etc.), having in mind a contribution to the understanding of the structure of electrolyte solutions, behavior of electrolytes in solution and, last but not least, supplying the scientific and technological communities with data on these important parameters in solution transport processes [3–7].

It is always convenient to keep in mind that all TP are macroscopic ones and that only with these data, it is not possible to understand the nature of the forces that bind the species, constituting these aqueous systems. Their interpretation can be made on the basis of models or theories. However, we would say that what normally is more important for the knowledge of the structure of the electrolyte solutions is the thermodynamic behavior of the involved species, not so much the complex question of the nature of their internal binding forces.

The search for information on the behavior and properties of aqueous solutions involving carbohydrates, ions, and drugs [16–18], has long been of great interest to many researchers due to their action in nature. This occurs, for example, in the biological processes of all organisms. However, their knowledge is still limited regarding the characterization of transport (diffusion, viscosity, and conductivity) and thermodynamic (density and activity coefficients) properties. Examples of such gaps include the electrolyte transport in biological systems caused by pH gradient, transport of ionic systems resulting from possible deterioration of dental materials in the oral cavity [19–26], as well as drug transport and possible complexes formed between them and cyclodextrins [27–30].

274 Green Chemistry and Green Engineering

In conclusion, we believe that the characterization of TP of aqueous systems containing electrolytes contributes to a better understanding of these systems, providing scientific and technological communities with some important parameter values for *in-vivo* pharmaceutical applications.

15.3.2 *VISCOSITY, CONDUCTIVITY, AND DIFFUSION: SOME CONCEPTS, TECHNIQUES, AND CRITICAL ANALYSIS OF DATA*

15.3.2.1 *VISCOSITY*

Among the TP, we have the viscosity (internal friction) that can be defined as the resistance against flow caused by an external force in liquids (and gases, or solids) (Eqn. 15.5) [3, 4].

$$F_{x,y} = \eta \frac{dv_x}{dy} \text{ (Newtonian Fluids)} \tag{15.4}$$

where $F_{x,y}$ is the force per unit of area required to move one plane relative to another, is the dynamic viscosity or viscosity (in centipoise, $\frac{dv_x}{dy}$ and is the velocity gradient.

Although it is convenient to define viscosity in terms of that hypothetical experiment, it is much easier to measure this property by other experimental processes. For example, it is possible to obtain measurements of viscosity by measurement of the flow velocity of a fluid through a cylindrical tube, and applying the Poiseuille equation (Eqn. 15.5):

$$\eta \frac{P\pi tr^2}{8Vl} \tag{15.5}$$

where V of liquid to flow through the capillary of length, l, and radius, r, under pressure, P, and t is the respectively time required during in this process. However, quantitative measurement of absolute viscosity by this process is difficult, being more frequent to measure relative viscosities of solutions, knowing previously the absolute viscosity of water.

Rotational viscometers are one of the more popular types of instruments used to measure the dynamic viscosity. This is the dynamic viscosity, represented by symbol η. The other process is to measure the resistive flow of a fluid under the weight of gravity. In this case, the kinematic viscosity

Transport Phenomena of Electrolytes in Aqueous Solutions 275

(v) is obtained. The capillary-based instruments are more accurate for its determining. The relation of these parameters (v and η) can be given by the following relation:

$$v \text{ (Kinematic viscosity) [in (cSt,)]} \times \rho \text{ (Density (in g cm}^{-3}) = \eta \text{ (Dynamic viscosity) [in (cP)]}$$

In some 1:1 electrolytes, we can trust the results of viscosity within ±0.1% (already much greater than the experimental error of 0.01%), but for many we cannot trust the results to much better than ±1% and in some cases, the uncertainties are greater than 1%. Polyvalent electrolytes: generally worst than 1%.

15.3.2.2   SPECIFIC CONDUCTIVITY

The resistance of a solution column of uniform cross-section with the area (cm²) between two inert electrodes of platine (Pt), representing d the distance between them, may be given by Eq. (15.6) [3, 4]:

$$R = \rho \frac{d}{A} \tag{15.6}$$

where ρ and A represent the specific resistivity and the area of theses electrodes. The inverse of the specific resistivity defines the specific conductivity, k (in ohm m^{-1}), that is:

$$k = \frac{1}{\rho} \tag{15.7}$$

As an example, the conductivities may be determined with the help of a three-electrode measuring cell, described elsewhere [31]. In this particular case, the cell was calibrated with dilute potassium chloride solutions. At the beginning of every measuring cycle, the cell was filled with a weighed amount of solvent. After measurement of the solvent conductivity at some temperature of the programme, a weighed amount of a stock solution was added using a gas-tight syringe (Figure 15.1).

Conductivity measurements in electrolyte solutions can be made with high accuracy and precision, certainly common error less than 0.01%. Resistance measurements can be made with an error of the order of 0.0001%, although we could not achieve this reproducibility from measure to measure in this laboratory.

FIGURE 15.1 Cell developed in Institutfür Physikalische und Theoretische Chemie, Regensburg.

Though conductivity can be measured with high precision (uncertainty less than ±0.01%), only in a few 1:1 electrolytes can be sure the results are right within 1.0–0.1%. In many others, the uncertainty is higher than 1%. Polyvalent electrolytes show still higher discrepancies: many electrolytes show discrepancies higher than 10%.

15.3.3 DIFFUSION

It is very common to find misunderstandings concerning the meaning of a parameter, called diffusion coefficient, in communications, meetings, scientific literature, and discussions with students or colleagues. Lobo et al. [32, 33], have clarified this concept. In fact, it is not a trivial topic, and, in addition, many papers reporting data on diffusion coefficients do not specify if they report to self or mutual diffusion. Having in mind these different concepts, it is convenient to define both highlight the differences between them.

Mutual diffusion is an important irreversible process taking place as a result of a gradient of concentration inside the solution. Its quantitative measure can be done through the diffusion coefficient of each species in solution. If there is concentration gradient in a binary electrolyte solution, then the constraint of maintaining electrical neutrality ensures that positive and negative ions move from the region of higher to lower concentration at the same speed. On a volume-fixed and most other reference frames, there is only one mutual or interdiffusion coefficient in this system. In this case, we need no theory for its interpretation, that is, we have a phenomenological interpretation: we base our thinking in the existence of a grad c which is an observable phenomenon and not a theory. There are not many cases where we can compare results from independent researchers [3, 4, 34, 35] (Figures 15.2 and 15.3). The author believes we can trust the results of Harned, Miller, Vitagliano, Leist, and Lobo with uncertainties of about 1%, but other sources from the literature are to be taken with care experimental methods that can be employed to determine mutual diffusion coefficients are diaphragm-cell (inaccuracy 0.5–1%), conductimetric (inaccuracy 0.2%), Gouy, and Rayleigh Interferometry (inaccuracy <0.1%) and Taylor dispersion (inaccuracy 1–2%) (Figure 15.3). While the first and second methods consume days in experimental time, the last ones imply just hours. Concerning the conductometric technique (Figure 15.2), despite this method has previously given us reasonably precise and accurate results, it is limited to studies of mutual diffusion in electrolyte solutions, and like diaphragm-cell experiments, the run times are inconveniently long (~days). The Gouy method also has high precision, but when applied to microemulsions they are prone to gravitational instabilities and convection. Thus, the Taylor dispersion has become increasingly popular for measuring diffusion in solutions, because of its experimental short time and its major application to the different systems (electrolytes or non-electrolytes). In addition, with this method, it is possible to measure multicomponent mutual diffusion coefficients.

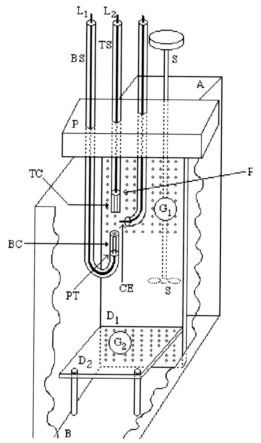

FIGURE 15.2 Scheme of the experimental set-up for the Lobo technique.

In self-diffusion the electrical neutrality, restraint does not apply and the three species (ions and solvent) have independent diffusion coefficients. (NMR and capillary-tube, the most popular methods, can only be used to measure intradiffusion coefficients).

We normally use theories to devise an acceptable interpretation: the atomic theory of matter and subsequently the molecular kinetic theory concerning matter. We cannot base our thinking on any macroscopic gradient of concentration.

There is no simple relation between mutual and self-diffusion coefficients except when the solute is at infinitesimal concentration. In this case, a variant of the Nernst equation can be written in its simplest form for 1:1 electrolytes as:

Transport Phenomena of Electrolytes in Aqueous Solutions

$$D_m^0 = 2\,(D_a^0 \cdot D_c^0)/(D_a^0 + D_c^0) \qquad (15.8)$$

where, D_m^0 is the limiting mutual diffusion coefficient of the salt and D_a^0 and D_c^0 are the limiting self-diffusion coefficients of anion and cation, respectively.

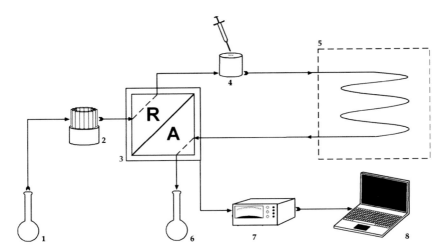

FIGURE 15.3 Scheme of the experimental set-up for the Taylor dispersion technique. (1) Carrier solution reservoir; (2) Metering pump (Gilson model Minipuls 3) to keep the flow-rate constant (0.17 mL min^{-1}); (3) Differential refractometer (Waters model 2410); (4) Injection 6-port valve (Rheodyne, model 5020); (5)Thermostatized airbox (298.15 ± 0.01) K) containing the dispersion tube; (6) Waste material; (7) Digital voltmeter (Agilent 34401 A; the voltage measurements were carried out at accurately 5 s intervals); (8) Computer (connected to the voltmeter through an IEEE interface). The flow direction is indicated by the arrows.

15.4 TRANSPORT PROPERTIES (TP) AND INFLUENCE OF TWO CONCEPTS: MEAN DISTANCE OF CLOSEST APPROACH OF IONS AND HYDRATION

Having in mind fundamental purposes, as well as the need in many technical fields, such as medicinal and pharmaceutical applications, we need to know accurate data concerning the fundamental thermodynamic and TP of solutions containing electrolytes. However, for the interpretation of those data with the application of models or theories, such as Onsager Fuoss equation to estimate diffusion coefficients, we need to know parameters such as the "mean distance of closest approach of ions," a (å when expressed in Angstroms). Despite considerable work has already been done, much of these data are not available from the literature, mainly due to the complexity involved in their estimation [36–41]. For example, this parameter, $a,$ depends not only on the nature of the electrolyte and its concentration, but also on the nature and concentration of the species present in the solution, which participate in the formation of an ionic atmosphere.

After some works, Lobo, and Ribeiro conclude that it is not possible to accurately know the mean distance of closest approach of ions, a, in an electrolyte solution, however desirable it would be. In fact, this physical meaning has been incompletely comprehended or is too complex to be mathematically analyzed. However, we present for different aqueous systems involving electrolytes, several estimations of a determined from different experimental techniques and/or theoretical approaches, so that the researcher who needs to use this parameter may have an idea of the possible range of values or, at least for same cases, two estimations for this parameter. All of them could be reasonable compromises to select an adequate a value, depending on their applicability to a given real problem. Consequently, by taking the appropriate precautions, each researcher can eventually either choose the most appropriate value for his case, or select a value from one specific method of estimation, or even use an average value of all of them or an average of the most suitable values for his case.

Thus, the applicability of these theoretical or semi-empirical equations is limited by the mathematical hypotheses and approximations on which the theory is based, and their results apply only to certain situations [3, 4]. For example, assuming complete dissociation of an electrolyte, the theory of Onsager Fuoss used on the estimation of D does not address the associated electrolytes, that is, weak electrolytes and those where an appreciable solute transport fraction occurs in the form of ionic pairs. A second limitation results from the non-contemplation of some effects which, although negligible for dilute solutions, are of great importance in the treatment of concentrated solutions. These include variations in viscosity, electrical permittivity, and number of hydration (h) [3, 43–45].

The interpretation of the number of hydration h is different depending of the nature of the property in the study. For example, in diffusion, and in the case of the salts, its interpretation as the number of water molecules moving with the ion as part of the diffusing unit may be accepted. However, from the treatment of activity coefficients data in terms of hydration, this parameter is defined as the effective number of molecules bound by the ion-solvent forces, and would therefore include contributions from water molecules beyond the first layer, which would not be firmly enough bound to move as a unit with the ion. Consequently, these latter values are higher when compared to those obtained from the treatment of diffusion coefficients data. However, the hydration numbers are somewhat lower than the majority of estimates of ionic hydration by other methods (RS).

In conclusion, no theory on TP in electrolyte solutions is capable of giving generally reliable data. The scarcity of these data in the scientific literature, due to the difficulty of their accurate experimental measurement and impracticability of their determination by theoretical procedures, allied to their industrial and research need, well justifies efforts in their accurate measurements.

ACKNOWLEDGMENTS

The authors in Coimbra are grateful for funding from "The Coimbra Chemistry Center" which is supported by the Fundação para a Ciência e a Tecnologia (FCT), Portuguese Agency for Scientific Research, through the programmes UID/QUI/UI0313/2019 and COMPETE. Ana P.C. Figueira is grateful to Prof. Antonietta Pinto of the University of Sapienza, Rome, Italy, for valuable comments about the study of metacomprehension.

KEYWORDS

- **conductivity**
- **diffusion**
- **electrolyte**
- **hydration**
- **mean distance of closest approach**
- **thermodynamics of irreversible process**
- **transport properties**
- **viscosity**

REFERENCES

1. Lobo, V. M. M., (1997). The definition of electrolyte: A comment from a reader and the author's reply. *Port. Electrochim. Acta, 15*, 215–219.

2. Lobo, V. M. M., (1996). The definition of electrolyte. *Port. Electrochim. Acta, 14*, 27–29.
3. Robinson, R. A., & Stokes, R. H., (1959). *Electrolyte Solutions* (2nd edn). Butterworths, London.
4. Tyrrell, H. J. V., & Harris, K. R., (1984). *Diffusion in Liquids* (2nd edn.), Butterworths, London.
5. Harned, H. S., & Owen, B. B., (1964). *The Physical Chemistry of Electrolytic Solutions* (3rd edn.). Reinhold Pub. Corp., New York.
6. Erdey-Grúz, T., (1974). *Transport Phenomena in Aqueous Solutions.* Adam Hilger, London.
7. Cussler, E. L., (1984). *Diffusion: Mass Transfer in Fluid Systems.* Cambridge University Press, Cambridge.
8. Lobo, V. M. M., (1990). *Handbook of Electrolyte Solutions.* Elsevier, Amsterdam.
9. Smith, E. B., (1973). *Basic Chemical Thermodynamics.* Oxford University Press.
10. Miller, D. G., (1956). Thermodynamic theory of irreversible processes. I: The basic macroscopic concepts. *Am. J. Phys., 24*, 433–436.
11. Miller, D. G., (1960). Thermodynamic of irreversible processes. The experimental verification of the Onsager reciprocal relations. *Chem. Rev., 60*, 15–37.
12. Miller, D. G., (1966). Application of irreversible thermodynamics to electrolyte solutions, I. Determination of ionic transport coefficients l_{ij} for isothermal vector transport processes in binary electrolyte systems. *J. Phys. Chem.,* 702639–702659.
13. Miller, D. G., (1967). Application of irreversible thermodynamics to electrolyte solutions. II. Ionic coefficients l_{ij} for isothermal vector transport processes in ternary electrolyte systems. *J. Phys. Chem., 71*, 616–632.
14. Miller, D. G., & Pikal, M. J., (1972). A test of the Onsager reciprocal relations and a discussion of the ionic isothermal vector transport coefficients l_{ij} for aqueous $AgNO_3$ at 25°C. *J. Sol. Chem.,* 1111–1130.
15. Miller, D. G., (1978). Ionic interactions in transport processes as described by irreversible thermodynamics. *Faraday Discuss. Chem. Soc., 64*, 295–301.
16. Horvath, A. L., (1985). *Handbook of Aqueous Electrolyte Solutions.* Physical properties. Estimation and correlation methods. John Wiley and Sons, New York.
17. Lobo, V. M. M., (1990). *Handbook of Electrolyte Solutions.* Elsevier Sci. Publ., Amsterdam.
18. Ribeiro, A. C. F., Lobo, V. M. M., Valente, A. J. M., Cabral, A. M. T. D. P. V., Veiga, F. J. B., Fangaia, S. I. G., Nicolau, P. M. G., et al., (2011). Transport properties and their impact on biological systems. In: Taylor, J. C., (ed.), *Advances in Chemistry Research* (Vol. 10, pp. 379–391). Nova Science Publishers. New York.
19. Koike, M., & Fujii, H., (2001). The corrosion resistance of pure titanium in organic acids. *Biomaterials, 22*, 2931–2936.
20. Schiff, N., Grosgogeat, B., Lissac, M., & Dalard, F., (2002). Influence of fluoride content and pH on the corrosion resistance of titanium and its alloys. *Biomaterials, 23*, 1995.
21. Alías, J. F. L., Gomis, J. M., Anglada, J. M., & Peraire, M., (2006). Ion release from dental casting alloys as assessed by a continuous flow system: Nutritional and toxicological implications. *Dent. Mater,* 22832–22837.
22. Upadhyay, D., Panchal, M. A., Dubey, R. S., & Srivastava, V. K., (2006). Corrosion of alloys used in dentistry. *Mater. Sci. Eng. A, 432*, 1–3.

Transport Phenomena of Electrolytes in Aqueous Solutions

23. Rezende, M. C. R. A., Alves, A. P. R., Codaro, E. N., & Dutra, C. A. M., (2007). Effect of commercial mouthwashes on the corrosion resistance of Ti-10Mo experiment alloy. *J. Mater. Sci. Mater Med., 18,* 149–154.

24. Rahm, E., Kunzmann, D., Döring, H., & Holze, R., (2007). Corrosion stable nickel and cobalt- based alloys for dental applications. *Microchim. Acta, 156,* 141–145.

25. Robin, A., & Meirelis, J. P., (2007). Influence of fluoride concentration and ph on corrosion behavior of titanium in artificial saliva. *J. Appl. Electrochem., 37,* 511–517.

26. Nascimento, M. L., Mueller, W. D., Carvalho, A. C., & Tomás, H., (2007). Electrochemical characterization of cobalt-based alloys using the mini cell system. *Dent. Mater.,* 23369–23387.

27. Del, V. E. M. M., (2004). Cyclodextrins and their uses: A review. *Process Biochem., 39,* 1033–1176.

28. Loftsson, T., & Duchêne, D., (2007). Cyclodextrins and their pharmaceutical applications. *Int. J. Pharm., 329,* 1–11.

29. Li, S., & Purdy, W. C., (1992). Cyclodextrins and their applications in analytical-chemistry. *Chem. Rev., 92,* 1457–1470.

30. Uekama, K., Hirayama, F., & Irie, T., (1998). Cyclodextrin drug carrier systems. *Chem. Rev., 98,* 2045–2076.

31. Barthel, J., Feuerlein, F., Neueder, R., & Wachter, R., (1980). Calibration of conductance cells at various temperatures. *J. Sol. Chem., 9,* 209–219.

32. Agar, J. N., & Lobo, V. M. M., (1975). Measurement of diffusion coefficients of electrolytes by a modified open-ended capillary method. *J. Chem. Soc., Faraday Trans., I 71,* 1659–1666.

33. Ribeiro, A. C. F., Natividade, J. J. S., & Esteso, M. A., (2010). Differential mutual diffusion coefficients of binary and ternary systems measured by the open-ended conductimetric capillary cell and by the Taylor technique. *J. Mol. Liquids, 156,* 58–64.

34. Callendar, R., & Leaist, D. G., (2006). Diffusion coefficients for binary, ternary and polydisperse solutions from peak-width analysis of Taylor dispersion profiles. *J. Sol. Chem., 35,* 353–379.

35. Barthel, J., Gores, H. J., Lohr, C. M., & Seidl, J. J., (1996). Taylor dispersion measurements at low electrolyte concentrations, I. Tetraalkyl ammonium perchlorate aqueous solutions. *J. Solut. Chem., 25,* 921–935.

36. Ribeiro, A. C. F., Esteso, M. A., Lobo, V. M. M., Burrows, H. D., Amado, A. M., Amorim, D. C. A. M., Sobral, A. J. F. N., et al., (2006). Mean distance of closest approach of ions: Sodium salts in aqueous solutions. *J. Mol. Liq., 128,* 134–139.

37. Ribeiro, A. C. F., Lobo, V. M. M., Burrows, H. D., Valente, A. J. M., Amado, A. M., Sobral, A. J. F. N., Teles, A. S. N., et al., (2008). Mean distance of closest approach of ions: Lithium salts in aqueous solutions. *J. Mol. Liq., 140,* 73–77.

38. Ribeiro, A. C. F., Lobo, V. M. M., Burrows, H. D., Valente, A. J. M., Sobral, A. J. F. N., Amado, A. M., Santos, C. I. A. V., & Esteso, M. A., (2009). Mean distance of closest approach of potassium, cesium, and rubidium ions in aqueous solutions: Experimental and theoretical calculations. *J. Mol. Liq., 146,* 69–73.

39. Ribeiro, A. C. F., Barros, M. C. F., Sobral, A. J. F. N., Lobo, V. M. M., & Esteso, M. A., (2010). Mean distance of closest approach of alkaline-earth metals ions in aqueous solutions: Experimental and theoretical calculations. *J. Mol. Liq., 156,* 124–127.

40. Kielland, J., (1937). Individual activity coefficients of ions in aqueous solutions. *J. Am. Chem. Soc., 59,* 1675–1678.

41. Marcus, Y., (1988). Ionic-radii in aqueous-solutions. *Chem. Rev., 88,* 1475–1498.
42. Lobo, V. M. M., & Ribeiro, A. C. F., (1994). Ionic association: Ion pairs. *Port. Electrochim. Acta, 12,* 29–41.
43. Ohtaki, H., & Radnai, T., (1993). Structure and dynamics of hydrated ions. *Chem. Rev., 93,* 1157–1204.
44. Lobo, V. M. M., & Ribeiro, A. C. F., (1995). Ionic solvation. *Port. Electrochim. Acta, 13,* 41–62.
45. Baes, C. F., & Mesmer, R. E., (1976). *The Hydrolysis of Cations.* John Wiley & Sons, New York.

Index

A

Acetic acid, 90, 97, 170
Acidification, 92, 103
Active site (AS), 168
Activity coefficient, 42
Acyclic molecules, 125
Adenosine
  triphosphoric acid, 246
  tri-phosphorus (ATP) acid, 250–253
Alkylamines, 152
Ambient temperature, 50, 65, 95, 96, 113, 138, 230
Amino acids, 58, 63, 184, 224
Ammonia, 24, 30, 31, 35–37, 39–42, 45, 174
  water cycle, 34
  water system, 36
Ammonium polyphosphate (APP), 243, 245, 248, 249, 253
Aneurysm, 13
Angiogenesis, 4
Antanopolous cycle, 38
Antibacterial
  activity, 200
  agents, 200
Antibiotic-resistant bacteria, 203
Anticancer drugs, 16, 84, 238
Antimalarial drug, 84
Anti-microbes, 245
Antimicrobial-antifungal activity, 14
Aquatic pollutants, 201
Aromatic
  compounds, 125
  molecules, 125, 126
  rings, 126
Ascorbic acid (AA), 166, 223, 227, 228, 230–238, 243, 245, 246, 250–253
Asymmetric molecules, 62
Asymptotic analysis, 124, 151
Atacama large millimeter array (ALMA), 184
Atom
  economy (AE), 85, 86, 88, 90, 104, 110, 113
  efficiency, 83, 88, 103

Atomic
  emission spectroscopy, 53
  energy, 57
  force microscopy (AFM), 68–70, 76, 80
    applications, 69
    theory, 69
  gases, 139
  magnetic moment, 244–248, 250, 252, 253
  mass spectrometry (AMS), 260, 267
  nuclei, 139
  radius, 125
  theory, 278
Atorvastatin, 98, 102
Attenuated total reflectance (ATR), 56, 57
  spectroscopy, 56
    applications, 57
    theory, 56
Auger
  effect, 58, 61
  electron spectroscopy (AES), 58, 80
    applications, 58
    theory, 58
Auxiliary agents, 86

B

Back-scattering geometry, 260
Bactericides, 245
Binary
  fluid, 24, 31, 45
  geothermal power plant, 27
  mixtures, 30
  power
    cycle, 43
    plants, 42
Bioactive
  agents, 1, 8
  substances, 243, 244, 247, 250
Bio-based polymers, 2
Biocatalysis, 78, 83, 95, 96, 103–105
Biocatalytic
  esterification, 104
  processes, 92, 96, 103, 104

reactions, 100
reduction, 98, 102
transformations, 102
Biochemistry, 73, 225
Biocomponents, 13
Biodegradable polymers, 13, 17
Biodegradation, 87
Bioengineering, 9
Biofibers, 2
Biogenic oxidative system, 228
Biological molecules, 70, 225
Biomacromolecules, 63, 64
Biomass, 25, 89, 97
Biomedical
applications, 8, 10, 17, 18, 19
fields, 14
sensing, 152
Biomedicine, 163, 179, 185
Biomimetic geometries, 13
Biomolecules, 8, 49, 50, 63–65, 70, 72, 73, 78, 80
inorganic interface, 65
interactions, 72
reactivity, 73
Bionanocomposites, 50, 57, 58, 64, 65, 70, 74, 75, 79, 80
materials, 49, 78
properties testing material properties, 74
mechanical properties, 74
optical properties, 74
Biopharmaceuticals, 97
Bioplastics, 2
Biopolymers, 9, 10, 75, 79
matrix, 75
Bioreactor, 200
Biosensors, 80
Biotechnology, 96, 199, 200
Biotransformations, 98
green matrices, 96
Bio-treatment, 102
Black hole (BH), 145
Bloch wave functions, 149, 150, 159
Block theorem, 157
Blood-brain barrier (BBB), 226
Boiling
point, 51, 52
water reactor (BWR), 213, 220
Bottoming cycle, 36, 37

Boundary motion model, 156
Brillouin zone (BZ), 157
Brownian motion, 71
Bulk fermentation, 97

C

Calorimetry, 64, 65, 164, 171
Cancer cells, 238
Carbohydrate, 142
Carbon
black filler location, 67
dioxide, 68, 94, 97, 116, 120, 132
efficiency (CE), 90, 164, 165
fibers, 244, 246, 248, 249
film, 68, 259
nanotubes, 59, 200, 201
shell, 244, 246, 248, 250
Carbonyl source, 132
Carbonylation, 93, 127
Cardiovascular disease, 13
Carnot
cycle, 45, 174
efficiency, 24, 30, 31, 34, 37, 39, 41
Cartography, 62
Cascading effect, 83, 94, 105
Catalysis, 83, 104, 105, 114, 127, 132, 174, 182, 224, 225
Catalytic systems, 164, 181
Cathode
active material, 153
materials, 153
reaction, 153
Cathodic reactions, 257
C-atoms, 125, 126
Cell
adhesion, 3, 14
attachment, 13, 14
bioconversion, 96
debris, 97
membrane, 79
migration, 13
surfaces, 69
Ceramic materials, 153
Ceylon graphite, 126
Chain
reactions (CRs), 140, 142
scission reactions, 52
Chaotic motion, 137

Index

287

Chelating agent, 64
Chemical
 bonds, 49, 59, 180, 185, 250
 engineering, 208
 science, 164, 170, 186
Chemocatalytic processes, 95
Chemoenzymatic process, 96
Chemophobia, 163, 165, 185
Chemotherapy, 17, 237
Chiral
 macromolecules, 63
 vector, 124, 151
Chitosan, 8, 14, 70
 chitin, 8
Chlorohydrin, 88, 99
Chlorophyll, 70, 142
Chromatography, 51, 164, 176, 177
Chronoamperometric (CA), 156
Circular
 dichroism (CD), 62, 63
  applications, 63
  spectroscopy, 63
  theory, 62
 economy (CE), 164
Citric acid, 97
Clotting
 activity, 11
 cloth, 11
Cluster, 78, 124, 127, 151, 244
 solvation models, 124, 151
Coagulation-fragmentation equations, 124, 151
Codexis, 98, 103
Coherent nature, 50
Cohesive force, 3
Cold plasma, 138, 139
Collagen
 interactions, 65
 triple helices, 65
Colloidal
 dispersion, 72
 solution, 72
 systems, 72
Colloids, 72
Conclusions (Cs), 9, 94, 146, 163, 179, 185, 202, 215
Conductivity, 6, 7, 29, 123, 124, 126, 131, 153, 175, 228, 269–273, 275, 276, 281
Conductometric technique, 277

Confrontation, 211, 215, 220
Constant phase element (CPE), 154
Coolidge plant, 29
Copolymers, 51, 67
Copper
 atomic magnetic moment, 246, 248
 carbon
  mesocomposite, 243–246, 248–250, 252, 253
  mesoscopic composite (Cu-C MC), 243, 245–248, 250, 253
Corona discharge, 138
Corrosion, 255–258, 260, 267
Couette cell, 77
Countercurrent chromatography (CCC), 176, 186
Critical
 point, 27, 67, 68
 temperature (CT), 128
Crucial juncture, 199
Cryo-electron microscopy, 62
Crystallization, 62, 169, 172
Curcumin, 70
Cyanation, 99, 102
 reaction, 99
Cycle power generation system, 24
Cyclic voltamperometer (CV), 154
Cyclodextrins, 273
Cysteine, 63, 237
Cytotoxicity, 14, 17, 79

D

Dark processes, 142
Degradation profile, 1, 9
De-polymerization, 51, 52
Dermal regrowth, 14
Design safe chemicals, 113
Diamagnetic anisotropy, 123, 131
Diamagnetism (DM), 124–126
Diazotizations, 93
Dicarboxylic acids, 51
Dictatorship, 220
Dielectric constant, 6
Diels-Alder reaction, 116
Diffraction, 53, 65, 245
Diffuse layer, 71, 72
Diffusion, 17, 71, 75, 153, 155–157, 269–274, 277–281
 coefficients, 277, 278

288

Index

Distillation, 34, 38, 41, 43, 44, 52, 94, 99, 102, 201
Drug
  carrier, 18
    ibuprofen, 17
  delivery, 1, 9, 14, 17, 19, 80
    applications, 13, 19
  diffusion, 17
  particles, 17
  releasing
    kinetics, 17
    rate kinetics, 2
    system, 17
Dynamic
  bioprocesses, 69
  light scattering (DLS), 65, 71, 72, 80
    theory, 71
  mechanical analysis (DMA), 75, 124–127
  viscosity, 274

E

E factor, 83, 85, 87–90, 92, 94, 96, 97, 101, 103, 104
Ecotoxicity, 92
Effective mass yield (EMY), 90, 105
Einstein's theory, 144, 145
Elastic
  domain, 76, 77
  modulus, 76, 79
Elastomer, 13
Electric
  conductance, 136
  field (EF), 4, 73, 78, 137, 138
  potential, 6
  power generation, 26
Electrochemical
  impedance spectroscopy (EIS), 154
  methods, 200
  reaction, 155
Electrochemistry, 156, 176
Electrode, 7, 73, 153–156, 160, 275
Electrodialysis, 201
Electrolysis, 92, 201
Electrolytes, 72, 140, 150, 153–156, 269–271, 273–281
Electromagnetic (EM), 74, 137, 249
Electromagnets, 78, 137

Electron, 53, 57–59, 61, 65, 125, 126, 136, 180, 185, 225, 244, 246, 248, 250, 269
  beam, 61, 62, 66
  energy loss spectroscopy (EELS), 61, 62
    applications, 62
    theory, 61
  microscopes, 65
  microscopy, 61, 62, 68, 78, 245, 246
  paramagnetic resonance (EPR), 223, 230, 231, 235–239, 245–249
  transparent replica, 68
  volts, 137
Electronic spin, 78
Electrophoresis, 73
  applications, 73
  theory, 73
Electrospinning, 1–9, 11–14, 17, 19
  equipment, 5, 6
  fundamentals, 3
  parameters, 3
  process, 4, 8, 9
  techniques, 9
Electrospray ionization (ESI), 150, 156, 160
Electrospun, 1–4, 7, 10, 11, 13, 14, 16–18
  nanofibers, 2, 3, 17
  polymer, 14
  scaffolds, 13
Electrostatic
  effect, 3
  properties, 72
  repulsion, 7
  spinning, 3
Elemental matrix effects, 60
Enantioselectivity, 86, 99
Endocytosis, 1, 8
Endothelial cell, 14
Endothermal reactions, 139
Energy
  challenges, 164, 170, 186
  dispersive x-ray spectroscopy (EDS), 60, 61, 261–264
    applications, 61
    theory, 61
  harnessing ways, 25
    binary power cycle, 30
    bottoming cycle applications, 30
    organic rankine cycle, 29
    salt-based power plants, 28
    solar thermal energy, 28

Index 289

solar thermal power, 29
supercritical cycles, 27
thermal power plants, 25
Enthalpy, 41, 42, 44, 64
Entropy, 42, 64, 124, 135, 143, 144, 146, 151, 271, 272
Environmental
engineering, 189–191, 193–196, 199, 201, 206–208
Protection Agency (EPA), 86
sustainability, 190, 191, 193, 196, 208
Equivalence principle, 145
Ethyl acetate solvents, 101
Eutrophication, 92
Excitonic effects, 129
Exergy, 24, 29, 31, 46
Extracellular
fluids, 11
matrix, 4, 10
Extraction flask, 51

F

Fascism, 212, 220
Fermentation, 89, 97
Fibrinogen, 8
Fibroblasts, 14, 15, 16
Field intensity, 78
Fine particle fraction (FPF), 172
Fluorescence (FLU), 59, 63, 78, 141
Fossil
energy, 170
fuel, 105, 168, 174
energy, 25
Fourier transform infrared spectrometry (FTIR), 54, 260
Franco
scheme, 211
system, 212, 219
Free radical (FRs), 139, 217, 224, 230, 235, 238
activity, 244, 250
Fresnel reflectors, 28
Frictional coefficient, 73
Friedel-Crafts acylations, 93
Fullerite crystal thermodynamic characteristics, 124, 151
Functional groups, 54, 55, 95, 244
Fundação para a Ciência e a Tecnologia (FCT), 281

G

Gas chromatography, 51, 52
Gaseous (G), 42, 68, 136
Gel matrix porosity, 73
Gelatin, 17
gel matrix, 75
Gene transfer, 203
General theory of relativity (GTR), 135, 146
Geofluid, 28
Geological effects, 60
Geothermal
fluid, 27, 28
power plant, 29
steam field, 29
Gibbs
free energy (G), 64
function, 135, 143, 146
Gibson Ashby model, 79
Glaxo Smith Kline (GSK), 89, 90
Glioblastoma tumor growth, 227
Global
positioning system (GPS), 144, 147
scenario, 200, 206
warming, 25, 92, 103, 119, 193, 195, 199, 203
Glucose dehydrogenase (GDH), 99, 100, 102
Glycosaminoglycan, 239
synthesis, 224
Gold-palladium nanocatalysts, 116
Grafting, 1, 9, 19, 62
Graphene (GR), 79, 124, 125, 127–130, 150–152, 157, 164, 175, 200, 201, 207
Graphite, 123–127, 131, 155, 156, 175, 260
Gravitational-wave (GW), 144
Green
chemistry, 83, 84, 86, 87, 89, 95, 100, 103, 109–113, 115–120
atom economy (AE), 87
atom efficiency, 101
benefits, 118
designing safer chemicals, 102
economy and business, 119
environment, 119
environmental impact factor (E factor), 87
greener solvents and auxiliaries, 102
human health, 118
matrices, 83, 87

nonhazardous chemical synthesis, 101
   principles applications, 100
   waste prevention, 100
engineering, 25, 46, 87
future, 120
matrices, 91
method, 113
metrics, 87, 90, 92, 104
principles, 109, 120
solvents, 115, 116
sustainability, 191, 193–195, 198, 199
sustainable energy, 25
synthesis, 111
technologies, 110, 111, 120
   principles, 120
Grignard reagent, 118

H

Halogenations, 93
Halogens, 59
Halohydrin dehalogenase (HHDH), 99, 100, 102
Hartree-Fock (HF), 93, 138, 139, 150, 154, 157
Hematopoiesis, 224
Hemostasis, 11, 12
Heteroatoms, 152
High
   frequency (HF), 138, 154
   temperatures (HTs), 126, 136, 219
Homeostasis, 13
Homogeneity, 128, 153
Human factor engineering, 206–208
Hyaluronan (HA), 14, 223, 227–239
Hybrid
   platforms, 70
   scaffolds, 9
Hybridization, 250, 252
Hydration, 279–281
Hydrocarbons (HCs), 29, 51, 94, 139, 147, 152
Hydrochloric acid, 270
Hydrodynamic radius, 71
Hydrogels, 75, 79
   mechanical testing, 75
Hydrogen peroxide, 88, 116
Hydrogenation, 93, 102
Hydrophilic interaction liquid chromatography (HILIC), 150

Hydrophilicity, 17, 226
Hydrophobic
   chains, 226
   tail, 167
Hydrophobicity, 65
Hydrophosphate ions, 258
Hydroxyapatite (HA), 14, 15
Hydroxyl radicals, 224
Hypotheses (H), 125, 136, 151, 165, 213

I

Ideal gas (IG), 137, 140
Imaging technique, 61, 65
Indentation techniques, 76
Inductively coupled plasma (ICP), 52, 53, 54
   advantages, 53
   atomic emission spectroscopy (ICPAES), 53, 60
   limitations, 54
Infrared (IR), 51, 54–56, 260, 264
   spectroscopy, 52, 56
Inorganic, 50
   biological interface, 49, 50, 80
   bioorganic interfacial systems, 72
   clusters, 74
   compounds, 54, 189
   materials, 151, 201
   organic interfaces, 58
   particles, 49, 78, 79
   salts, 88, 93, 97
Integrated water resource management, 191, 192, 195, 196, 198, 202, 204, 206, 208
Intensification, 83, 105
Inter-band electronic transitions, 62
Interfacial adhesion, 79
Intergalactic medium, 164, 184
Intermolecular forces, 256, 267
Interstellar space, 182, 183
Intradiffusion coefficients, 278
In-vitro
   evolution, 96, 99
   studies, 8, 18
In-vivo
   pharmaceutical applications, 274
   studies, 8, 13, 14, 18
Ionic
   liquids, 94, 116
   strength, 72, 79, 151

Index

291

Ion-solvent forces, 280
Irradiation, 56, 57
Isobutane, 27, 29
Isoelectric point, 72, 74
Isothermal
  plasma, 137
  titration calorimetry (ITC), 64, 65, 78
    applications, 64
    determining size, 65
    structure imaging, 65
    theory, 64
Isotropic system, 272

J

James Webb Space Telescope (JWST), 184
Josephson junctions, 78
Juxtaposition, 194

K

Kalina cycle, 24, 30–32, 37, 38, 46
Keratinocyte, 14
Keto ester, 99
Ketones, 51
Ketoreductase (KRED), 99, 100, 102
Key questions (KQs), 177
Kinematic viscosity, 274
Kinetic energy, 57, 61, 136

L

Layered double hydroxides (LDHs), 124
Leucine groups, 63
Lewis acids, 93
Life cycle
  analysis (LCA), 83, 92, 103, 104
  assessment, 83, 92, 105, 114
Light
  elements, 61
  emission, 61
Linear equations, 272
Lipids, 70, 224
Liquid (L), 3, 30, 33, 34, 36–38, 40–42, 46, 52, 54, 56, 67–69, 72, 74, 77, 117, 118, 136, 137, 150, 176, 255, 270, 274
  liquid biphasic media, 94
  mole fraction, 42
Live cell imaging, 69
Lorentz efficiency, 24

Lotka-Volterra equations, 212, 213, 219
Low
  k dielectric, 160
  pressure (LP), 27, 34, 37–39, 44, 45, 117, 137

M

Macrocosmos, 135, 146
Macromolecular
  complexes, 64
  properties, 51
Macromolecule, 63, 229
Macroscopic
  properties, 270
  quantity gradient, 270
Macrostructures, 177
Magnetic
  anisotropy (MA), 123–129, 131
    effect (MAE), 129
  fibers, 7
  field (MF), 6, 7, 78, 137, 139, 140, 244, 245
    force lines (MFLs), 137
    trap (MT), 137
  groups, 149, 150, 159
  measurements, 77
  mesoparticles, 244, 245
  mesoscopic particles, 245, 252
  metal-carbon mesoscopic composites, 250
  moment, 245, 248, 250, 252
  particle clusters, 74
  properties, 123, 124, 126, 131, 192, 250, 252
  susceptibilities (MSs), 51, 53, 123–126, 131, 150, 176
Magnetochemistry, 132
Magnetohydrodynamic (MHD), 139
Magnetoresistance, 129
Malae cycle, 41
Malony Robertson cycle, 31
Manganese
  porphyrins (MnPs), 224–226, 228, 230, 233, 235, 237–239
  superoxide dismutase (MnSOD), 224
Mass
  fraction, 34, 37, 38
  intensity (MI), 90
  productivity (MP), 26, 34, 38, 90, 137, 138
  spectrometry, 186, 260, 267

Fourier transform infrared spectroscopy (MS-FTIR), 184
spectroscopy, 53
Matrices, 54, 73, 79, 83, 84, 90, 104, 105, 155, 177
Matrix effect, 176
Maxwell
Faraday equation, 78
law, 136
Medium pressure (MP), 26
Mesenchymal stem cells, 14
Mesocomposite (MC), 243–245, 247–250, 252, 253
Mesoparticle, 244, 245, 250, 252
Mesoporous
materials, 150, 151
silica material, 160
Mesoscopic
composites, 244, 247, 253
copper carbon composite, 253
particle, 246
Metallic conduction, 123, 124, 126, 131
Metalloenzymes, 224
Metalloporphyrins, 224
Metalloproteins, 224
Metal-organic frameworks (MOFs), 125, 151
Methylene chloride, 51
Microbial
growth, 13
hydroxylation, 84
Microelectronics, 175
Microemulsions, 277
Microfabrication, 192
Microfluids, 14
Microliter drop, 68
Micrometers, 2
Micromolar levels, 225
Micropollutants, 199
Microscopic techniques, 66
electron microscopy, 66
theory, 66
Microstructures, 18, 177
Microtoming, 67
Microwave (MWs), 117, 164, 177, 231
Millennium development goals (MDG), 203
Mobil composition material (MCM), 151, 152
Molar mass, 227, 229, 230, 238, 239
Mole fraction, 14, 15, 31, 39, 40, 42

Molecular
complexity, 84
conformation, 73
crystals, 124, 125
electronics, 160
interactions, 64
kinetic theory, 278
magnetism, 128
mass, 73, 74
migration, 73
radius, 125
sieve, 152, 168
weight, 5, 51, 88
Molten salt, 28, 29
Monkeyman, 212, 213, 220
Monochromatic light, 53
Monochromator, 53, 260
Monolayers, 129
Monomer, 52
Mori Tanaka models, 79
Multicomponent
fluid, 36
mixture, 24, 46
Multidimensional chromatography, 176
Multi-temperature boiling, 24
Mutual diffusion, 277

N

N-acetylcysteine (NAC), 237
Nananocomposite materials, 49
Nanoadsorbents, 199–201, 207
Nanocomposites (NCs), 14, 15, 49, 59, 79, 80, 150–152
Nanocones, 124, 151
Nanoengineering, 192, 194, 195, 197
Nanofibers, 1–11, 13–19
Nanofibrous mats, 9, 11, 12, 13
Nanofillers, 2, 79
Nanomaterials (NMs), 62, 152, 190–197, 199–202, 205–208
Nanometers, 2, 3, 192
Nanoparticles, 9, 62, 63, 70, 74, 78, 79, 116, 168, 203, 204
Nanoscale (NS), 69, 70, 74, 123, 124, 130, 132, 192, 201
Nanoscience, 8, 116, 189
Nanostructures, 9, 18, 68, 245
Nanotechnology, 1, 8, 19, 129, 189–202, 206–208

Natural gas (NG), 139, 166
Nazism, 220
Near ambient pressure (NAP), 58
Neoorgans, 13
Neotissues, 13
Nernst equation, 278
Neutral
  molecules, 136
  particles, 138
Neutron scattering, 70
Non-equilibrium thermodynamics, 271
Non-renewable
  energy, 25
  sources, 25
Non-solar-based energy, 25
Novel
  cell-biomaterial interfaces, 13
  cycle, 45
Nuclear
  fuels, 25
  fusion, 136, 213, 218
  magnetic resonance (NMR), 51, 62, 167,
    278, 167
    analysis, 62
  physics, 136, 213
  science, 136, 212, 213, 219
  weapons, 136, 212, 213
Nucleic acids, 63, 74
Nyquist plot, 154, 157

O

Ohmic drops, 155
Oleochemistry, 118
Oligo-nucleotides, 70
Onsager Fuoss equation, 279
Oral cavity, 273
Organic
  fluids, 29
  liquids, 60, 153
  Rankine cycles (ORC), 29, 46
  reactions, 116
  reagents, 93
  solvent, 94, 117, 124, 151
    dispersions, 124, 151
Organo-catalysis, 104
Organo-metallic substances, 51
Osteogenic capacity, 14
Oxidation
  reactions, 247

states (OSs), 123, 124, 128, 131, 132, 159,
  245
Oxidative stress, 79, 224–226

P

Paramagnetism (PM), 124, 126
Partial condensation, 36
Particle
  biomolecule interactions investigation, 72
  physics, 136, 144, 147
  suspension, 77
Periodic table, 123, 124, 132, 149–151, 157
  of the elements (PTE), 123, 124, 129,
    131, 132, 149, 150, 159
Periodicity, 180, 185
Perkin-Elmer model, 260
Petrochemicals, 2, 97
pH, 17, 58, 63, 72, 95–97, 99, 102, 103,
  115, 151, 152, 170, 257, 273
Pharmaceutical, 89, 92–96, 116, 201
  drugs, 119
  industries, 94
Phase transitions (PTs), 75, 155
Phonons, 62
Phosphate glass, 262, 265
Phosphoric acid, 98, 182, 243, 245, 257,
  261–264, 267
Phosphorus acid esters, 51
Photo/electro reduction, 127, 132
Photocatalysis, 199, 203
Photocatalytic
  applications, 203
  production (PCP), 164, 175
Photochemical reactions, 135, 141, 142, 146
Photochemistry, 118, 120, 141
Photodiode, 68, 69
Photoelectric effect, 57
Photoelectrons, 57, 59
Photoionization, 58
Photomultiplier tube (PMT), 53
Photon correlation spectroscopy (PCS), 71
Photovoltaic (PV), 25, 171
Plasma (P), 52, 53, 60, 65, 135–140, 147,
  212, 213, 219
  application, 147
  column, 137
  generation, 147
  particles, 137

Plasmochemistry, 139
Plasmotrons, 139, 140
Plasticity, 76, 77
Plasticizers, 50, 51
Platine (Pt), 59, 176, 275
Plethora, 49, 127
Poiseuille equation, 274
Polarity, 3, 117
Poly lactic-co-glycolic acid (PLGA), 14, 16
Poly(ethylene oxide) (PEO), 7, 152
Poly(propylene oxide) (PPO), 152
Polyacetylene, 244, 246, 253
Polyacrylamide, 73
Polyatomic molecules, 125
Polyesters, 8
Polyglycolide (PGA), 8
Polylactic acid (PLA), 8, 14, 16
Polymer, 7, 8, 14, 51, 59, 60, 66, 67, 75, 118, 199
  characterization, 52
  chemical analysis, 50
  foams, 79
  granules, 5
  identification, 50, 51
  mechanical comminution, 50
  molecules, 79
  qualitative analysis, 51
  quantitative analysis, 51
Polymeric
  electrospun nanofibers healthcare applications, 12
    drug delivery, 16
    tissue engineering, 12
    wound dressing, 14
  filaments, 3
  materials, 18
  solution, 3, 9, 11
Polymerization, 140
Polymorphism, 171
Polyoxometalate (POM), 124, 125, 127, 151
Polyvalent electrolytes, 275, 276
Polyvinyl alcohol, 243, 245
Porosity, 1, 11, 75, 153
Porous materials (PMs), 124–126, 151, 153
Potassium dichromate, 84
Power
  cycle, 26, 28, 29, 44, 45, 46
  sectors, 25

Prebiotic molecules, 182, 183
Pregabalin synthesis, 96
Primary fluid, 34, 36
Progesterone, 84
Proliferation, 9, 11, 13, 14, 17, 79, 227
Pro-oxidative effect, 227, 230, 232, 234, 235, 238
Protonation, 57
Pyrolysis, 51

Q

Qualitative
  analysis, 51, 52, 60, 67
  chemical analysis, 61
  development, 114
  inorganic analysis, 51
Quantitative
  analysis, 54, 59, 60, 61
  development, 114
Quantum
  equivalence, 141
  mechanics (QM), 135, 146, 147
  yield, 135, 142, 143, 146
Quartz
  cuvettes, 74
  electrodes, 6, 7
Quasi-Cottrellian approach, 156

R

Radical processes inhibitors, 253
Radio frequency waves, 53
Radiochemical reactions, 135, 136, 140, 142, 143, 146, 213
Radiography, 212
Radiolysis, 142
Radioprotective effects, 225
Rankine
  cycle, 23–30, 37, 45, 46
  steam cycle, 30
Rapid evaporation, 68
Raw materials, 85–87, 94, 102, 109, 113, 118, 120, 166
Reaction
  mass efficiency (RME), 90, 105
  time, 181, 186
Reactive oxygen
  and nitrogen species (ROS/RNS), 225
  species, 224, 239

Real-time analysis, 114
Reboiler temperature, 44, 45
Recuperator, 30, 36
Redlich
  Kister equation, 42
  Kwong equation, 42
Redox catalysts, 224
Reductants, 85, 225, 226
Reduction-oxidation, 245
Reflux, 41, 43
Refractive index, 52, 65
Refractometer, 279
Refrigerants, 29
Relative humidity, 3, 75, 258
Relaxation tests, 77
Remediation, 86, 87, 102, 119, 189–191,
  194–197, 199–201, 205–208
Renewable energy (RE), 25, 158, 169
Residence time, 181, 186
Rheology, 77
Robertson cycle, 30
Rogdakis cycle, 39
Room temperature (RT), 64, 78, 99, 102,
  123, 124, 126, 131, 230, 260
Rotational viscometry, 223, 235, 239
Rust conversion, 255, 256, 260, 267
  layers (RCL), 260, 263, 264, 267

S

Safety storage cabinet, 164, 172, 173, 186
Santa Barbara Amorphous (SBA), 152
Scaffold, 1, 2, 9–11, 13–19, 80
  fabrications, 9
Scanning
  electron microscopy (SEM), 15, 61,
    66–68, 76, 259, 261–264, 267
    samples preparation, 67
  probe microscopy (SPM), 69
  transmission electron microscopy
    (STEM), 62
  tunneling microscopy (STM), 69
Scattering
  angle, 70
  techniques, 70
    applications, 71
    theory, 70
  vector (Q), 70
Sedimentation, 201

Self-consistent field (SCF), 150
Semiconductors, 74
Silica
  nanoparticles, 63
  particles, 65
Silk fibroin (SF), 14, 15
Silver
  nanoparticles, 70
  staining, 73
Simulation analysis, 43
Single-wall
  carbon nanotubes (SWNTs), 124, 151
  nature of nanohorns (SWNHs), 124, 151
Solar
  energy, 23–26, 28, 29, 45
  fuels, 164, 175
    photocatalytic production, 164, 175
  panels, 28
  steams, 45
  thermal panel, 28
Solid (S), 8, 52, 54, 56–58, 66, 74, 75, 77,
  79, 85, 93, 96, 136, 137, 149, 150, 153,
  155, 157, 159, 270
Solvent
  casting, 9
  etching, 66
Soxhlet apparatus, 51
Space-time yield, 94, 100
Specific surface area (SSA), 151, 192
Spectrometric methods, 60
Spectroscopic
  methods, 63
  techniques, 52
    inductively coupled plasma (ICP), 52
Spin orientation, 129
Stabilizers
  qualitative analysis, 52
  quantitative analysis, 52
Standard models (SMs), 135, 146
Statistical chain-scission reactions, 52
Steam
  cycle, 36
  power
    cycle, 34
    plant, 23, 24, 28
Stereoselectivity, 95
Stoichiometric
  amounts, 88
  equation, 88, 90

quantitative data, 65
quantities, 85, 93
reagents, 87, 89, 95, 114
Stoichiometry, 64, 85, 90, 118
Stokes-Einstein equation, 71
Sulfuric acid, 85, 92, 97, 270
Superacids, 124, 151
Supercapacitor, 160
Superconducting quantum interference
  device (SQUID), 78
Superconductors (SCs), 123, 124, 128, 129,
  131
Supercritical
  carbon dioxide, 94
  cycle, 27, 28
  fluid, 117
  nuclei, 156
  operation, 28
Superheater, 30, 41, 44
Surface
  analysis techniques, 58
  plasmon resonance (SPR), 65, 78
  saturation, 65
  tension, 3, 4, 6, 67, 68
  to volume ratio (SVR), 4, 8, 11, 68
Sustainable
  chemistry, 112, 114, 115, 120, 167
  development goals (SDG), 204
  reaction media, 93
  route, 1, 9, 19

T

Tensile
  strength, 75
  tests, 76, 77
Tetraethylorthosilicate (TEOS), 152
Therapeutic
  agents, 1, 18, 19
  applications, 19, 237
  compounds, 16
  effects, 17
Therapeutically active substances, 243–246,
  250, 252, 253
Thermal
  conductivity, 29
  efficiency, 24, 26, 30, 31, 34, 37, 39–41,
    44
  power plant, 25, 28

Thermodynamic, 23, 26, 63–65, 124, 139,
  151, 270–273, 279, 281
  changes, 64
  data, 63
  of irreversible processes (TIP), 270–273,
    281
  parameters, 65
Thermograms, 52
Thermonuclear
  plant, 140
  reactions, 139
  reactors, 140
  synthesis, 139
Thin
  films (TFs), 66, 76, 153
  layer chromatography (TLC), 51
Tip-to-collector (TCD), 3
Tissue
  engineering, 9, 12, 14, 18, 152
  regenerations, 1, 9, 13, 18, 19
Topography, 10, 11, 57, 68, 69
Topological
  materials
    periodic table, 157
  quantum chemistry (TQC), 149, 150,
    157–159
Toxic
  chemical, 118, 119
  wastes, 119
Traditional steam power plants, 23
Transesterification, 51
Transition
  analytic electron microscope, 246
  electron microscopy with high permission
    (TEM HP), 246
  metal, 78, 116, 224, 227, 231, 239
    dichalcogenides (TMDs), 124, 128
Transmission electron microscopy (TEM),
  61, 62, 65–68, 245–248
Transmittance, 55
Transport properties (TP), 269–271, 273,
  274, 279, 281
Trinitrotoluene (TNT), 85
Tris(triazolyl)methane, 182
Turbine inlet, 41, 44, 46
  pressure, 44
Turbulence dynamics, 219
Twisted bilayer GR, 128

Index

297

Two-dimensional (2D), 124, 125, 127–130, 132, 149–151, 158, 159, 177
    material physics, 132

U

Ultrafiltration, 201
Ultrasound sonication, 117
Ultraviolet (UV), 52, 56, 74, 118, 141
Unpaired electrons, 78, 244, 246, 248–250, 253
Unsaturated hydrocarbons (UHCs), 139
Urban management scenario, 205
Urotropine, 243, 245, 246, 250–253
UV-Vis spectrophotometry, 78

V

Vacuum, 58, 65, 99, 102, 135–137, 141, 145–147, 184, 246, 259
Vapor
    generator, 38, 39
    liquid
        equilibrium, 42, 46
        separator, 37
    mass fraction, 37
Vascular tissues, 13
Vasoconstriction, 11
Velocity, 7, 73, 139, 274
Vibrating sample magnetometer (VSM), 78
Vickers hardness test, 76
Virial equation, 42
Viscosity, 3, 5, 11, 71, 73, 231–235, 269, 271, 273–275, 280, 281
Visible (VIS), 56, 89, 118, 141, 235
Vital force, 84

W

Wastewater
    systems, 197
    treatment, 97, 190, 191, 195, 198–200, 206, 207
Water
    pollutants, 198

quality management, 192, 196, 198, 202 205, 208
resource management, 191, 194, 196, 201–206
Wavelength, 53, 56, 55, 59, 62, 65, 70, 74, 141, 143
    dispersive x-ray fluorescence spectrometry (WDXRF), 59
Weissberger's biogenic oxidative system (WBOS), 227, 230–233, 235, 236, 238
Wistar rat model, 18
Wohler's synthesis, 84
World War I (WWI), 212–216, 220
Wound dressings, 9, 19, 80

X

X-ray
    beam, 59
    crystallography, 62
    diffraction, 62
    diffractograms, 265
    fluorescence spectrometry (XRF), 59, 60
        applications, 59
        theory, 59
    line spectrum, 59
    photoelectron
        spectra, 252
        spectroscopic investigations, 246
        spectroscopy (XPS), 57–59, 61, 245
Xylene, 53

Y

Young modulus, 75, 76

Z

Zerovalent ion, 201
Zeta potential, 65, 72
Zetametry, 71, 72
    applications, 72
    theory, 72
Zinc oxide (ZnO), 64, 70, 75